Lecture Notes in Control and Information Sciences

Volume 490

This series reports new developments in the fields of control and information sciences—quickly, informally and at a high level. The type of material considered for publication includes:

1. Preliminary drafts of monographs and advanced textbooks
2. Lectures on a new field, or presenting a new angle on a classical field
3. Research reports
4. Reports of meetings, provided they are
   (a) of exceptional interest and
   (b) devoted to a specific topic. The timeliness of subject material is very important.

Indexed by EI-Compendex, SCOPUS, Ulrich's, MathSciNet, Current Index to Statistics, Current Mathematical Publications, Mathematical Reviews, IngentaConnect, MetaPress and Springerlink.

More information about this series at https://link.springer.com/bookseries/642

Ashton Roza · Manfredi Maggiore · Luca Scardovi

Distributed Coordination Theory for Robot Teams

Springer

Ashton Roza
Department of Electrical and Computer Engineering
University of Toronto
Toronto, ON, Canada

Manfredi Maggiore
Department of Electrical and Computer Engineering
University of Toronto
Toronto, ON, Canada

Luca Scardovi
Department of Electrical and Computer Engineering
University of Toronto
Toronto, ON, Canada

ISSN 0170-8643 ISSN 1610-7411 (electronic)
Lecture Notes in Control and Information Sciences
ISBN 978-3-030-96086-5 ISBN 978-3-030-96087-2 (eBook)
https://doi.org/10.1007/978-3-030-96087-2

This Springer imprint is published by the registered company Springer Nature Switzerland AG
The registered company address is: Gewerbestrasse 11, 6330 Cham, Switzerland

To Joyce Erskine

—Ashton Roza

To Sachiko and Emilia

—Manfredi Maggiore

To Lucia, Gastone, Priama, and Kaya-Marisa

—Luca Scardovi

Contents

Notation and Abbreviations

$(v_1, \ldots, v_n)$ in $(\mathbb{R})^n$	Column vectors $v = [v_1 \cdots v_n]^\mathrm{T}$ in $\mathbb{R}^n$ are identified with n-tuples $(v_1, \ldots, v_n)$ in $(\mathbb{R})^n$
$v \cdot w := v^\top w$	Euclidean inner product of $v, w \in \mathbb{R}^n$
$\\|v\\| := (v \cdot v)^{1/2}$	The Euclidean norm of v
$\mathbb{R}_+$	The set of positive real numbers
$\{e_1, e_2\}$	The natural basis of $\mathbb{R}^2$
$\{e_1, e_2, e_3\}$	The natural basis of $\mathbb{R}^3$
$\mathbf{1}$	The vector of ones in $\mathbb{R}^n$
$\otimes$	The Kronecker product of matrices
$\mathbb{S}^1$	The unit circle, which we identify with the set of real numbers modulo 2π
$\mathbb{S}^n$	The n-dimensional unit sphere
$\mathbb{T}^n$	The n-torus $\mathbb{T}^n := \mathbb{S}^1 \times \cdots \times \mathbb{S}^1$ (n times)
$A \backslash B$	The set-theoretic difference of the sets A and B
$\bar{A}$	The closure of the set A
$\lvert A\rvert$	The cardinality of the set A
$B_\varepsilon(\Gamma)$	The ε neighborhood of the closed set Γ
$N(\Gamma)$	A neighborhood of the closed set Γ
$\mathbf{n}$	$\{1, \ldots, n\}$
$\mathbf{k} : \mathbf{n}$	$\{k, \ldots, n\}$
$(x_j)_{j \in I}$	If $I = \{i_1, \ldots, i_n\}$ is an index set, the ordered list of elements $(x_{i_1}, \ldots, x_{i_n})$ is denoted by $(x_j)_{j \in I}$
$L_f \varphi(x)$	If $f(x)$ is a vector field and $\varphi(x)$ is a C^1 Function $x \mapsto L_f \varphi(x) := (d/dx)\varphi(x) \cdot f(x)$ is the Lie derivative of φ along f

Notation for Kinematic Unicycle Model

$x_i \in \mathbb{R}^2$	Inertial position of unicycle i
$x \in \mathbb{R}^{2n}$	$(x_i)_{i\in\mathbf{n}}$
$R_i \in SO(2)$	Attitude of unicycle i
$\theta_i \in \mathbb{S}^1$	Heading angle of unicycle i
$\theta \in \mathbb{T}^n$	$(\theta_i)_{i\in\mathbf{n}}$
$\omega_i \in \mathbb{R}$	Angular velocity of unicycle i
$r^i = R_i^{-1} r$	Coordinate representation of r in frame B_i
$x_{ij} = x_j - x_i$	Relative displacement of robot j with respect to Robot i
$\theta_{ij} = \theta_j - \theta_i$	Relative heading angle of robot j with respect to Robot i
N_i	Set of neighbors of robot i
$y_i = (x_{ij})_{j\in N_i}$	Vector of relative positions available to robot i
$\varphi_i = (\theta_{ij})_{j\in N_i}$	Vector of relative angles available to robot i

Notation for Flying Robots

m_i, J_i	Mass and inertia matrix of robot i
$x_i \in \mathbb{R}^3$	Inertial position of robot i
$x \in \mathbb{R}^{3n}$	$(x_i)_{i\in\mathbf{n}}$
$v_i \in \mathbb{R}^3$	Linear velocity of robot i
$v \in \mathbb{R}^{3n}$	$(v_i)_{i\in\mathbf{n}}$
$R_i \in SO(3)$	Attitude of robot i
$\Omega_i \in \mathbb{R}^3$	Angular velocity of robot i
$\Omega \in \mathbb{R}^{3n}$	$(\Omega_i)_{i\in\mathbf{n}}$
$q_i = -R_i e_3$	Thrust direction vector of robot i
$T_i = -u_i R_i e_3$	Applied thrust vector of robot i
$r^i = R_i^{-1} r$	Coordinate representation of r in frame B_i
$x_{ij} = x_j - x_i$	Relative displacement of robot j with respect to Robot i
$v_{ij} = v_j - v_i$	Relative velocity of robot j with respect to Robot i
$\Omega_i \in \mathbb{R}^3$	Reference angular velocity of robot i
N_i	Set of neighbors of robot i
$y_i = (x_{ij}, v_{ij})_{j\in N_i}$	Vector of relative positions and velocities available to robot i

Notation in Chaps. 7 and 8

$\mathcal{C}$	Path to follow
$c^\star(x)$	Orthogonal projection of x onto the line path $\mathcal{C}$
$d_{1i}^1 \in \mathbb{R}^2$	Desired displacement of robot i with respect to robot 1
$\bar{v} > 0$	Desired path following speed

$c \in \mathbb{R}^2$	Circle center for path following
$p_i \in \mathbb{S}^1$	Desired heading offset of robot i with respect to robot 1
$p \in \mathbb{R}^2$	Beacon
$\theta_p \in \mathbb{S}^1$	Angle of p
$\delta_i \in \mathbb{R}^2$	Offset vector attached to frame B_i
α_i	Component of δ_i parallel to $R_i e_1$
β_i	Component of δ_i perpendicular to $R_i e_1$
$\widehat{x}_i = x_i + \delta_i$ or $\widehat{x}_i = x_i + \beta_i R_i e_2$	Offset position of unicycle i, in the case of final linear motion or final circular motion, respectively
$f_i((x_{ij})_{j \in N_i})$	Integrator consensus controller
$g_i((\theta_{ij})_{j \in N_i}, \eta)$	Rotational integrator consensus controller
$h(x)$	Integrator path following controller

Abbreviations

The following is a list of abbreviations for the coordination problems presented in this book. See Chap. 3 for the problem formulations.

Abbreviation	Explanation	Target set
RP-F	**R**endezvous **P**roblem for **F**lying Robots	Γ
RP-U	**R**endezvous **P**roblem for Kinematic **U**nicycles	Γ
PP	**P**arallel Formation **P**roblem	Γ_p
PFP	**P**arallel Formation **F**locking **P**roblem	Γ_{pf}
PFP-B	**P**arallel Formation **F**locking **P**roblem with a **B**eacon	Γ_{pfb}
PPP	**P**arallel Formation **P**ath Following **P**roblem	Γ_{lp}
CFP	**C**ircular Formation **F**locking **P**roblem	Γ_{cf}
CPP	**C**ircular Formation **P**ath Following **P**roblem	Γ_{cp}

Next, the list of abbreviations for concepts of set stability. These are presented in Appendix A.

Abbreviation	Where	Explanation
LAS	Definition A.7	**L**ocal **A**symptotic **S**tability
GAS	Definition A.7	**G**lobal **A**symptotic **S**tability
AGAS	Definition A.10	**A**lmost-**G**lobal **A**symptotic **S**tability
SGAS	Definition A.11	**S**emi-**G**lobal **A**symptotic **S**tability
ASGAS	Definition A.12	**A**lmost **S**emi-**G**lobal **A**symptotic **S**tability
GPS	Definition A.13	**G**lobal **P**ractical **S**tability

Chapter 1
Introduction

Abstract This chapter introduces the notion of distributed coordination in the field of robotics. This concept is illustrated through a number of historical and modern examples. We then provide an overview of the coordination problems investigated in this manuscript.

1.1 Motivation

In this monograph, we present algorithms for the coordination of autonomous teams of robots in order to achieve a desired collective goal using limited sensor data. Coordination problems in robotics have been the object of scientific interest for a long time. In a letter written in 1665 to the Royal Society of London, the Dutch physicist Christiaan Huygens noted that two pendulum clocks mounted on a beam have a natural tendency to synchronize, in that they oscillate at the same frequency. He observed that the synchronizing mechanism was a minute oscillation of the beam that would effectively couple the pendulum dynamics inducing synchronization. Figure 1.1 contains the original drawing by Huygens describing the pendulum apparatus. In modern terminology, Huygens described a phenomenon of *synchronization* on the torus, and the coupling mechanism of the beam is referred to as a synchronizing feedback.

In 1732, French mathematician and hydrographer Pierre Bouguer presented a paper at the French Academy in which he studied so-called pursuit curves traced by one ship chasing another. The question was whether the two ships would collide. This idea has resurfaced many times in a variety of forms in the intervening centuries eventually giving rise to the cyclic pursuit problem [1, 2]: N kinematic point robots on the plane pursue each other in a cyclic fashion, with agent i pursuing agent $i+1$. Will the agents converge to each other? The answer is yes, they will converge to the average of their initial positions. This cyclic pursuit idea is the simplest *rendezvous* algorithm.

In 1987, Reynolds [3] wrote a computer program called *boids* (android birds) simulating a flock of birds flying with a common average heading and avoiding collisions. Motivated by this work, Vicsek et al. in 1995 [4] formalized the Boids algorithm in a discrete-time setting modeling a collection of particles with the same

A. Roza et al., *Distributed Coordination Theory for Robot Teams*,
Lecture Notes in Control and Information Sciences 490,
https://doi.org/10.1007/978-3-030-96087-2_1

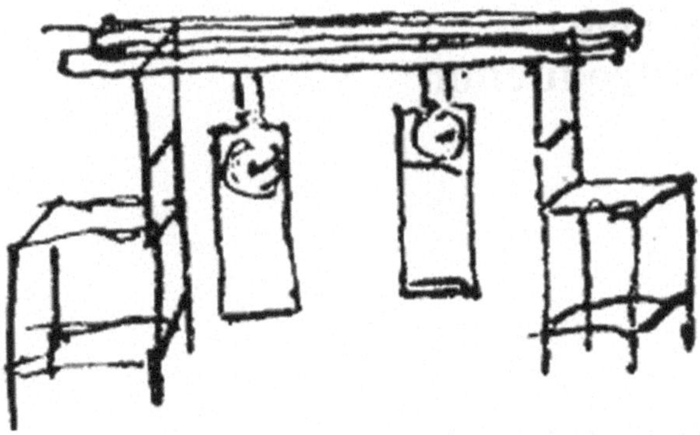

Fig. 1.1 Original drawing by C. Huygens describing the pendulum apparatus of his 1665 synchronization experiment

speed and whose heading is updated using a local averaging rule. We call the update rule of Vicsek et al. a *flocking* algorithm. The stability of this algorithm was analyzed by Jadbabaie et al. in [5].[1]

The above are just some of the historical roots of modern distributed coordination theory. Further literature is reviewed in Chap. 10. The problems highlighted above have two characteristics in common. There is a group of agents, and each agent takes decisions based on relative information measured with respect to nearby agents it can see, but without having any information about its own state or the absolute state of its neighbors. A control algorithm possessing this characteristic is called *local and distributed* (more about this in Chap. 2). Further, using such distributed decision-making, the collection of agents exhibits some form of *collective goal* (synchronization, rendezvous and flocking in the examples above).

With recent technological advances in the area of robotics, distributed coordination problems have become viable and crucially important. Consider, for example, the setup of Fig. 1.2. A team of autonomous quadrotor helicopters equipped with on-board cameras is given the task of monitoring a forested area during the dry season, searching for any sign of fire. The helicopters must organize an efficient formation that follows a desired path at a desired velocity such that the entire dry area is scanned. This could be achieved by flying in a parallel fashion between a series of straight line segments, circling about each way-point with a desired radius and for a desired time duration. Mathematically, the problem just described is called *formation path following*.

As another example, one may like to have formations of robots follow a series of linear and circular path segments painted or taped on the floor of a warehouse to retrieve and stock merchandise.

As a final example, consider a team of automatic snowplows with the task of snow removal on highways. Typically, the snowplows form a diagonal line formation, as illustrated in Fig. 1.3, passing accumulated snow from one plow to the other until it clears to the shoulder of the highway. In Fig. 1.3, snow is being passed from vehicle

[1] The setup of [4] and the theoretical results of [5] turn out to be special cases of the so-called agreement algorithm that preceded their development, see, e.g., [6].

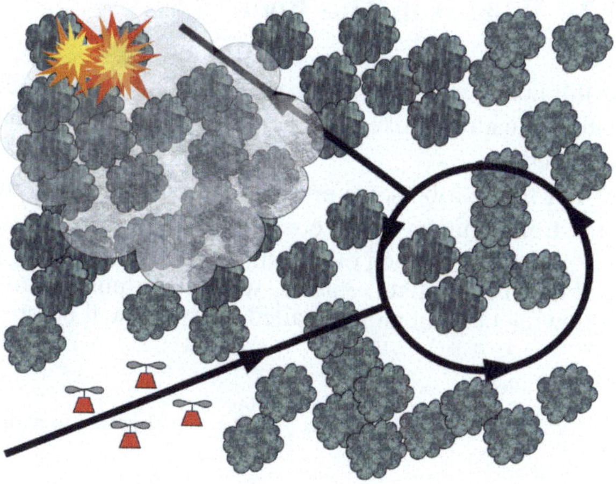

Fig. 1.2 Quadrotor fire inspection. The center of the formation moves along the path in the direction indicated

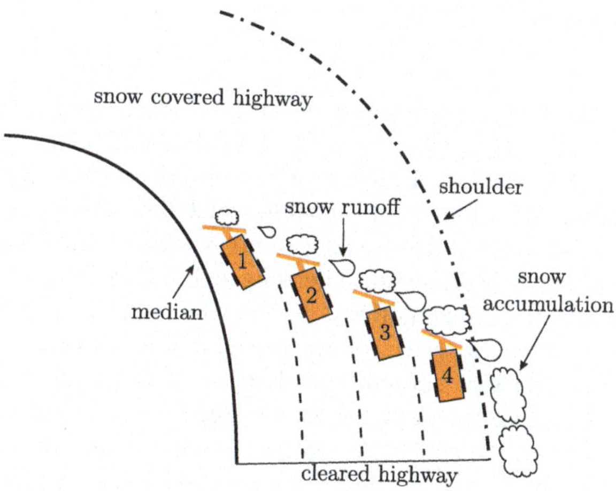

Fig. 1.3 Example of snow removal on a highway

1 to vehicle 4 before clearing the highway. Each plow only observes the vehicle to its left and right in the process. These are all examples of distributed coordination problems.

This monograph aims to provide a systematic framework for formulating distributed coordination problems of this kind and new ideas for solving them. A unifying viewpoint of this text is the formulation of all basic coordination problems in terms of the stabilization of certain subsets of the state space of the ensemble of agents. We propose a hierarchical perspective whereby some of the more complex control specifications are broken down into a hierarchy of simpler specifications that we call *distributed control primitives*. Relying on this point of view, this text presents a number of control algorithms representing the first solutions to certain coordination problems that have so far eluded researchers.

1.2 What Is in This Book

In this book, we solve a number of coordination problems for two important classes of robots: quadrotor-like underactuated flying robots and unicycle-like nonholonomic vehicles.

For flying robots, we present the first solution to the following rendezvous problem: from arbitrary initial conditions, make the robots convene to an arbitrarily small neighborhood of each other using only camera information.

For unicycle-like vehicles, we solve a number of problems summarized in the following table (a more detailed overview will be presented in Chap. 3).

Collective goal / Final motion	Rendezvous	Full sync	Formation path following	Formation flocking
Rest	×	×		×
Parallel		×	×	×
Circular		×	×	×

Each problem in the table above is characterized by two attributes: which collective goal is to be achieved and which final motion is desired. When the collective goal is rendezvous, i.e., make all robots convene to some common location, a controller is presented achieving this goal in which the vehicles come to rest.

When the collective goal is full synchronization, i.e., make all robots converge to each other with identical heading, controllers are presented achieving this goal with a variety of final motions: ones in which the vehicles come to rest, follow some line not defined a priori, follow a specific line defined a priori, or move along a circle. It is clear that a controller solving the full synchronization problem also solves the rendezvous problem, but we will keep two such solutions distinct.

When the collective goal is formation path following, i.e., make robots arrange themselves in a desired formation and follow a path given a priori, solutions are presented for paths that are straight lines and circles.

Finally, when the collective goal is formation flocking, i.e., make robots arrange themselves in a desired formation and move as a group, solutions are presented in which the group motion is static, along a straight line or along a circle. While the type of motion here (static, linear or circular) is decided beforehand, unlike in formation path following, no specific path to be followed is defined a priori.

As we mentioned earlier, all problems investigated in this monograph are formulated in terms of the stabilization of a suitable subset of the collective state space of the robots, and controllers are designed using basic modules that we call distributed control primitives.

In most problems, the solutions we present are local and distributed in the sense made precise in Chap. 2, but which, roughly speaking, means that to implement the proposed algorithms, each robot relies on relative information acquired through observation of a subset of its neighbors. No communication is required between robots, nor a GPS device or an external camera system. For the formation path following controllers, naturally the algorithms require the measurement of the relative distance of the robot to the path.

The monograph has been written under the assumption that the reader has taken a basic undergraduate/graduate course in state-space linear systems theory and an introductory graduate-level course on nonlinear systems. Appendices A and B contain background material on stability theory and graph theory that will bring the reader up to speed with the tools required to understand the problem formulations and proofs. In particular, Appendix A presents precise definitions of various notions of set stability, and among other things, it reviews an essential tool, a so-called reduction theorem for asymptotic stability of sets. Additionally, a new reduction theorem for almost global asymptotic stability is developed.

1.3 What Is Not in This Book

Distributed coordination theory is a developing area with a number of open problems. A key challenge in this area is the sensor requirements, namely the restriction that feedbacks should be local and distributed in the sense defined in Chap. 2. While this book presents the first solution of the rendezvous problem for underactuated flying robots like quadrotor helicopters, it does not present any formation control algorithm for such aircraft, and indeed, the formation control problem is still open, as there are currently no local and distributed solutions to it.

In this monograph, as is standard in the literature, the interconnection structure between robots is modeled by a visibility graph, and this graph is assumed to be static. This assumption is questionable from a practical point of view, but it is convenient to present a complete theoretical analysis of various algorithms. In particular, we do not consider time-varying or state-dependent sensor graphs. In Chap. 9, however, we present extensive simulations analyzing the performance of certain algorithms when the sensor graphs are state dependent (this occurs, typically, when each agent can only see a certain number of its closest neighbors).

Collision avoidance is an important requirement in coordination problems, but it is not considered in this monograph. The reason, once again, is that many of the problems treated in this book were open until now, and it is natural to first solve existing open problems before adding the collision avoidance requirement. We hope that this text will help spur research aimed at improving our algorithms with a collision avoidance component.

Finally, all presented algorithms require the measurement of relative quantities such as relative positions and orientations. This can be done using markers on each robot and on-board cameras, but the image processing algorithms required to convert image data to relative quantities are not presented in this text.

1.4 Book Organization

This book is organized as follows:

- Chapter 2: Robot Models
  In this chapter, we present models for kinematic unicycles and flying robots and precisely define what it means for a feedback controller to be local and distributed.
- Chapter 3: Coordination Problems
  This chapter presents the distributed coordination problems treated in this monograph, and it formulates them precisely in terms of stability properties of a suitable subset of the collective state space of the robots.
- Chapter 4: Distributed Control Primitives
  This chapter reviews the fundamental building blocks that are used to solve the complex coordination problems in this monograph. These are rendezvous controllers for single and double integrators, synchronization controllers for rotational integrators and attitude synchronization controllers for systems described by rotation matrices.
- Chapter 5: Rendezvous of Flying Robots
  This chapter presents a local and distributed feedback for the rendezvous of underactuated flying vehicles. This is the first solution to this problem.
- Chapter 6: Rendezvous of Kinematic Unicycles by Means of Local and Distributed Feedback
  This chapter presents a local and distributed solution to the rendezvous problem for kinematic unicycles. An important feature of this solution is that the feedback is smooth and time independent. Previous solutions to this problem require either time-varying or discontinuous feedback or are restricted to undirected sensing graphs.
- Chapter 7: Unicycle Formations Coming to Rest
  From the rendezvous problem, in this chapter, we move on to the problem of stabilizing formations of kinematic unicycles with the requirement that, asymptotically, the unicycles come to a stop. This is a stepping stone towards more complex coordination problems investigated in the following chapters. A special case of this problem is when the unicycles fully synchronize; i.e., both positions and orientations converge to each other. This is qualitatively different than the rendezvous controllers of Chap. 6 that only make the positions of the unicycles converge and not their orientations. The controller presented in this chapter is the first to solve the stated problem.
- Chapter 8: Unicycle Formations with Parallel and Circular Motions
  This chapter continues the investigation of the distributed formation control problem, focusing on the collective goals of path following and flocking, and final motions that are either parallel or circular. This makes for a total of four coordination problems. Concerning the formation flocking problem with parallel final motions, we also consider the case when a beacon is available that gives a common sense of direction (such as a compass), and in this case, we obtain a solution valid for a general class of sensor digraphs containing a globally reachable node,

whereas without a beacon, our solution is only guaranteed to work for a class of hierarchical sensor graphs. Almost all solutions presented in this chapter are the first solutions of the associated problems.

- Chapter 9: Unicycle Formation Simulation Trials
This chapter presents extensive simulation trials to study the effectiveness of our control solution presented in Chap. 7 for formation control of unicycles under different realistic scenarios not captured by the main theoretical result in that chapter. This includes performance in the presence of undirected state-dependent sensor graphs, relaxation of high gain requirements and robustness to unmodeled effects including sensor noise, input noise, sampling and saturated inputs.

References

1. Lin, Z., Broucke, M., Francis, B.: Local control strategies for groups of mobile autonomous agents. IEEE Trans. Autom. Control **49**(4), 622–629 (2004)
2. Marshall, J.A., Broucke, M.E., Francis, B.A.: Pursuit formations of unicycles. Automatica **2**(1), 3–12 (2006)
3. Reynolds, C.W.: Flocks, herds and schools: a distributed behavioral model. In: Proceedings of the 14th Annual Conference on Computer Graphics and Interactive Techniques (SIGGRAPH '87), vol. 21, pp. 25–34 (1987)
4. Vicsek, T., Czirók, A., Ben-Jacob, E., Cohen, I., Shochet, O.: Novel type of phase transitions in a system of self-driven particles. Phys. Rev. Lett. **75**(6), 1226–1229 (1995)
5. Jadbabaie, A., Lin, J., Morse, A.S.: Coordination of groups of mobile autonomous agents using nearest neighbor rules. IEEE Trans. Autom. Control **48**(6), 988–1001 (2003)
6. Tsitsiklis, J.N., Bertsekas, D.P., Athans, M.: Distributed asynchronous deterministic and stochastic gradient optimization algorithms. IEEE Trans. Autom. Control **31**(9), 803–812 (1986)

Chapter 2
Robot Models

Abstract In this chapter, we present models of kinematic unicycles and flying robots and give a precise definition of *local and distributed feedback*.

2.1 Attitude Representation and Coordinate Frames

In preparation for modeling, in this section, we discuss rotation matrices and their geometric representation in terms of orthogonal coordinate frames. To begin, we introduce the *special orthogonal group* $\mathsf{SO}(k)$, $k \in \{2, 3\}$, the set of matrices

$$\mathsf{SO}(k) = \{R \in \mathbb{R}^{k\times k} : R^\top R = I_k = RR^\top, \det(R) = 1\},$$

where I_k is the $k \times k$ identity matrix. An element R of $\mathsf{SO}(k)$ is called a *rotation matrix*. As its name suggests, $\mathsf{SO}(k)$ constitutes a matrix group under matrix multiplication.

Rotation matrices are in a one-to-one relationship with orthogonal coordinate frames. To explore this relationship, we start with the observation that, by definition, rotation matrices have the property that $R^\top R = I_k$, and this identity is equivalent to the property that each column of R is a unit vector, and all columns of R are mutually orthogonal. If we represent the columns of R as unit vectors attached at a common origin, R is identified with an orthogonal frame in $\mathbb{R}^k$ whose origin is not fixed. The property $\det(R) = 1$ implies that this frame is right-handed. In summary, rotation matrices in $\mathsf{SO}(k)$ are geometrically represented by right-handed orthogonal frames in $\mathbb{R}^k$, as claimed.

Now consider a team of n robots and two right-handed orthogonal frames, $\mathcal{I}$ and $\mathcal{B}_i$, either in $\mathbb{R}^2$ (Fig. 2.1) or $\mathbb{R}^3$ (Fig. 2.2). Suppose that $\mathcal{I}$ is an inertial frame, while $\mathcal{B}_i$ is attached to the ith agent in the team, with $i \in \{1, \ldots, n\}$. The *attitude* of robot i

---

A. Roza et al., *Distributed Coordination Theory for Robot Teams*, Lecture Notes in Control and Information Sciences 490,
https://doi.org/10.1007/978-3-030-96087-2_2

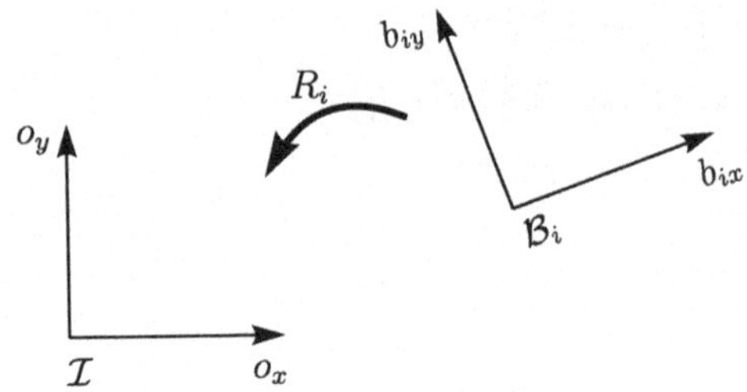

Fig. 2.1 Inertial and body frames in two dimensions

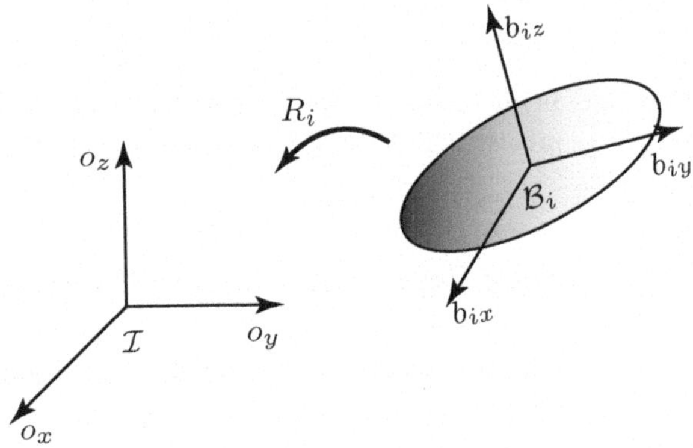

Fig. 2.2 Inertial and body frames in three dimensions

is defined to be the rotation matrix of frame $\mathcal{B}_i$ expressed in the coordinates of frame $\mathcal{I}$, i.e.,

$$
\begin{aligned}
R_i &:= \begin{bmatrix} b_{ix} \cdot o_x & b_{iy} \cdot o_x \\ b_{ix} \cdot o_y & b_{iy} \cdot o_y \end{bmatrix}, \\
R_i &:= \begin{bmatrix} b_{ix} \cdot o_x & b_{iy} \cdot o_x & b_{iz} \cdot o_x \\ b_{ix} \cdot o_y & b_{iy} \cdot o_y & b_{iz} \cdot o_y \\ b_{ix} \cdot o_z & b_{iy} \cdot o_z & b_{iz} \cdot o_z \end{bmatrix}
\end{aligned}
\tag{2.1}
$$

for the cases of $\mathbb{R}^2$ and $\mathbb{R}^3$, respectively, where "$\cdot$" denotes the dot product of two geometric vectors. One can check that R_i as defined in the first equation of (2.1) is in $\mathsf{SO}(2)$ while the one in the second is in $\mathsf{SO}(3)$.

Besides representing the attitude of robot i, the matrix R_i can be used to change coordinate representations of vectors. If v^i are the coordinates of a vector in frame $\mathcal{B}_i$, then one can show that the coordinate representation of the same vector in frame $\mathcal{I}$ is $v = R_i v^i$. In this book, we adopt the convention that a vector with a superscript i is represented in the coordinates of frame $\mathcal{B}_i$. If a vector does not have any superscript, then it is represented in frame $\mathcal{I}$. Thus, in the above, v^i is a body frame vector in frame $\mathcal{B}_i$, and $v = R_i v^i$ is the same vector represented in frame $\mathcal{I}$.

Matrices $R_i \in \mathsf{SO}(2)$ can always be represented in terms of angles as

$$
R_i = \begin{bmatrix} \cos(\theta_i) & -\sin(\theta_i) \\ \sin(\theta_i) & \cos(\theta_i) \end{bmatrix}
$$

for a suitable angle $\theta_i \in \mathbb{S}^1$ (the set of real numbers modulo 2π, which can be identified with the unit circle). Indeed, with reference to (2.1), we see that θ_i is the angle between the geometric vectors o_x and b_{ix}, measured counterclockwise from o_x. Owing to the bijective relationship between angles in $\mathbb{S}^1$ and rotation matrices in $\mathsf{SO}(2)$, the matrix group $\mathsf{SO}(2)$ can be identified with the unit circle $\mathbb{S}^1$ (more precisely, they are diffeomorphic manifolds).

Now suppose agent i moves in space so that its rotation matrix depends on time, $R_i = R_i(t)$. Suppose, as is natural, that all entries of $R_i(t)$ are C^1 functions. We now show that R_i can be viewed as the solution of a special matrix differential equation. For simplicity, we begin with the two-dimensional case. We have seen that we may represent $R_i(t)$ as

$$R_i(t) = \begin{bmatrix} \cos(\theta_i(t)) & -\sin(\theta_i(t)) \\ \sin(\theta_i(t)) & \cos(\theta_i(t)) \end{bmatrix}.$$

Differentiating with respect to t, one obtains

$$\begin{aligned} \dot{R}_i &= \begin{bmatrix} -\sin(\theta_i(t)) & -\cos(\theta_i(t)) \\ \cos(\theta_i(t)) & -\sin(\theta_i(t)) \end{bmatrix} \dot{\theta}_i(t) \\ &= R_i \begin{bmatrix} 0 & -\dot{\theta}_i(t) \\ \dot{\theta}_i(t) & 0 \end{bmatrix}. \end{aligned}$$

Let $\omega_i(t) := \dot{\theta}_i(t)$, and denote

$$\omega^\times = \begin{bmatrix} 0 & -\omega \\ \omega & 0 \end{bmatrix}.$$

Then,

$$\dot{R}_i = R_i\, \omega_i(t)^\times. \tag{2.2}$$

This is the kinematic equation of a rotating body in $\mathbb{R}^2$, and $\omega_i(t) = \dot{\theta}_i(t)$ is called the *angular speed* of the body.

A similar differential equation holds for rotations in $\mathbb{R}^3$. If $R_i(t) \in \mathsf{SO}(3)$, then differentiating the identity $R_i(t)^\top R_i(t) = I_3$ one can show that $R_i{}^\top \dot{R}_i$ is skew symmetric. Since all skew symmetric matrices 3×3 have the form

$$\Omega^\times := \begin{bmatrix} 0 & -\Omega_3 & \Omega_2 \\ \Omega_3 & 0 & -\Omega_1 \\ -\Omega_2 & \Omega_1 & 0 \end{bmatrix},$$

for some $\Omega \in \mathbb{R}^3$, there exists a unique continuous function $\Omega_i^i : \mathbb{R} \to \mathbb{R}^3$ such that $R_i{}^\top \dot{R}_i = (\Omega_i^i(t))^\times$, or

$$\dot{R}_i = R_i\, \Omega_i^i(t)^\times. \tag{2.3}$$

The vector $\Omega_i^i(t)$ is the *angular velocity of agent i with respect to frame $\mathcal{I}$, referenced in body frame $\mathcal{B}_i$*, and Eq. (2.3) is the kinematic equation of a rotating body in $\mathbb{R}^3$. In the vector Ω_i^i, the subscript denotes the robot and the superscript denotes the frame in which the vector is referenced, as we discussed earlier.

2.2 Models

Having reviewed the rotation matrix parametrization of attitude in two and three dimensions, we are ready to introduce the two vehicle classes investigated in this monograph. The first class corresponds to ground-based nonholonomic mobile robots modeled as kinematic unicycles. The second class corresponds to underactuated flying robots.

2.2.1 *Kinematic Unicycles*

A kinematic unicycle model is illustrated in Fig. 2.3. This figure also depicts a differential-drive robot which fits the model class. We fix an orthogonal frame $\mathcal{I} = \{o_x, o_y\}$ in $\mathbb{R}^2$ and attach to unicycle i an orthogonal body frame $\mathcal{B}_i = \{b_{ix}, b_{iy}\}$ in such a way that b_{ix} is the heading axis of the unicycle (as in Fig. 2.1). We denote by $x_i \in \mathbb{R}^2$ the position of unicycle i in the coordinates of frame $\mathcal{I}$. The attitude of body frame $\mathcal{B}_i$ relative to $\mathcal{I}$ is represented by a rotation matrix $R_i \in \mathsf{SO}(2)$. Unicycle i's wheels point in the direction of its heading axis $R_i e_1$ which represents the instantaneous axis of motion for unicycle i. This means that the unicycle velocity, $\dot{x}_i$, is proportional to the unit vector $R_i e_1$, and letting $u_i \in \mathbb{R}$ denote the linear speed of the unicycle, we thus have $\dot{x}_i = u_i R_i e_1$. Letting $\omega_i \in \mathbb{R}$ denote the angular speed of the unicycle, we have seen that $\dot{R}_i = R_i \omega_i^\times$. Putting everything together, we get

$$\begin{aligned} \dot{x}_i &= u_i R_i e_1, \\ \dot{R}_i &= R_i \omega_i^\times. \end{aligned} \tag{2.4}$$

This is the model of a kinematic unicycle. The pair (u_i, ω_i), composed of the linear and angular speeds of unicycle i, is the control input. The state, the pair $(x_i, R_i) \in \mathbb{R}^2 \times \mathsf{SO}(2)$, can be grouped into the homogeneous transformation matrix

$$H_i = \begin{bmatrix} R_i & x_i \\ 0_{1\times 2} & 1 \end{bmatrix}.$$

The set of all such matrices forms a group under matrix multiplication and is denoted $\mathsf{SE}(2)$, the *Special Euclidean group*. Thus the unicycle state space is $\mathsf{SE}(2)$.

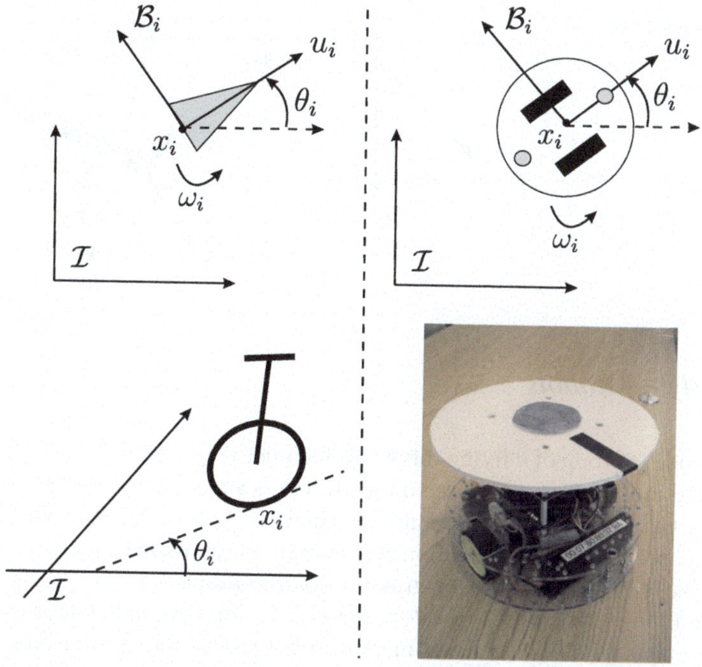

Fig. 2.3 Kinematic unicycle class. The figure on the right shows a differential-drive robot that is modeled as a unicycle

There is an equivalent representation of the unicycle dynamics. Recall that, letting $\theta_i \in \mathbb{S}^1$ be the angle between vectors o_x and b_{ix}, one can write the rotation matrix R_i in terms of θ_i as

$$R_i = R_i(\theta_i) = \begin{bmatrix} \cos(\theta_i) & -\sin(\theta_i) \\ \sin(\theta_i) & \cos(\theta_i) \end{bmatrix},$$

and in this case, the angular speed is given by $\omega_i = \dot{\theta}_i$. Using this fact, we may rewrite (2.4) as

$$\begin{aligned} \dot{x}_i &= u_i R_i e_1 = \begin{bmatrix} \cos(\theta_i) \\ \sin(\theta_i) \end{bmatrix} u_i \\ \dot{\theta}_i &= \omega_i . \end{aligned} \tag{2.5}$$

The state of (2.5) is the pair $(x_i, \theta_i) \in \mathbb{R}^2 \times \mathbb{S}^1$. The state spaces $\mathsf{SE}(2)$ and $\mathbb{R}^2 \times \mathbb{S}^1$ are diffeomorphic and the models (2.4) and (2.5) are equivalent.

We collect the translational and rotational states into the vectors $x := (x_i)_{i \in \mathbf{n}} \in \mathbb{R}^{2n}$ and $\theta := (\theta_i)_{i \in \mathbf{n}} \in \mathbb{T}^n$. The overall system state space is $\mathbb{R}^{2n} \times \mathbb{T}^n$.

Fig. 2.4 Flying robot class

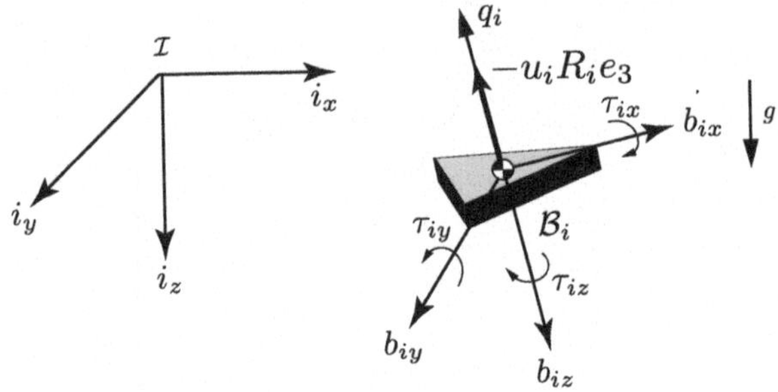

2.2.2 *Flying Robots*

We now model a group of n flying robots. Referring to Fig. 2.4, we fix a right-handed orthonormal inertial frame $\mathcal{I}$, common to all robots and attach at the center of mass of robot i a right-handed orthonormal body frame $\mathcal{B}_i = \{b_{ix}, b_{iy}, b_{iz}\}$. We denote by (x_i, v_i) the coordinates of the position and velocity of robot i measured in the inertial frame $\mathcal{I}$. We let $g \in \mathbb{R}^3$ denote the gravity vector in frame $\mathcal{I}$.

For the flying robot class, we adopt the convention that an actuator produces a thrust force applied at the center of mass of robot i. This thrust force can be varied in intensity, but its direction is constant in body frame and assumed to be directed opposite to the body frame axis b_{iz}, as depicted in Fig. 2.4. Letting $q_i := -R_i e_3$, a unit vector, the thrust force is given by $u_i q_i$, where $u_i \in \mathbb{R}$ is a control input. We refer to q_i as the *thrust direction vector* of robot i. By Newton's third law, the translation of vehicle i is governed by the equation

$$\ddot{x}_i = -u_i R_i e_3 + m_i g.$$

Now, we turn to the attitude dynamics. To begin with, denoting by $R_i \in \mathsf{SO}(3)$ the attitude of body frame $\mathcal{B}_i$ relative to $\mathcal{I}$, we have the kinematic equation

$$\dot{R}_i = R_i (\Omega_i^i)^\times,$$

where recall from Sect. 2.1, $\Omega_i^i \in \mathbb{R}^3$ is the angular velocity of robot i with respect to frame $\mathcal{I}$, referenced in body frame $\mathcal{B}_i$. Now we need to determine the dynamics of Ω_i^i. To this end, we make the assumption that the robot has an actuation mechanism inducing control torques $\tau_{ix}, \tau_{iy}, \tau_{iz}$ about its body axes. Letting $\tau_i := (\tau_{ix}, \tau_{iy}, \tau_{iz})$ be the torque vector, the angular velocity dynamics read as

$$J_i \dot{\Omega}_i^i = \tau_i - \Omega_i^i \times J_i \Omega_i^i.$$

Picking $(x_i, v_i, R_i, \Omega_i^i)$ as the state for robot i, we obtain the equations of motion

$$\begin{aligned}\dot{x}_i &= v_i,\\ m_i\dot{v}_i &= -u_i R_i e_3 + m_i g = T_i + m_i g,\end{aligned} \tag{2.6}$$

$$\begin{aligned}\dot{R}_i &= R_i\,(\Omega_i^i)^\times,\\ J_i\dot{\Omega}_i^i &= \tau_i - \Omega_i^i \times J_i\Omega_i^i.\end{aligned} \tag{2.7}$$

In the above, m_i is the mass of robot i and $J_i = J_i^\top$ is its inertia matrix. The overall state space of (2.6), (2.7) is $\mathbb{R}^{3n} \times \mathbb{R}^{3n} \times \mathsf{SO}(3)^n \times \mathbb{R}^{3n}$.

The model (2.6), (2.7) is standard and is widely used in the literature to model flying vehicles such as quadrotor helicopters. See, for instance, [3]. Sometimes, researchers use alternative attitude representations, such as quaternions [4] or Euler angles [5, 6]. The model (2.6), (2.7) ignores aerodynamic effects such as drag and wind disturbances (such effects are included in [3]). It also ignores the dynamics of the actuators.

Example 2.1 A well-known vehicle that falls in the class of flying robots is the quadrotor helicopter, see [5] or [7]. Referring to Fig. 2.5, a quadrotor helicopter consists of four rotors connected to a rigid frame. The distance from the center of mass to the rotors is denoted by d. For robot i, each rotor produces a thrust force f_{ij}, $j \in \{1, \dots, 4\}$ parallel to the b_{iz} axis, and a reaction torque τ_{rij} of the motor that drives it. To produce a thrust f_{ij} in the negative b_{iz} (i.e., upward) direction, the two rotors on the b_{ix} axis rotate in the clockwise direction, while the rotors on the b_{iy} axis rotate in the counterclockwise direction.

The physical inputs are the reaction torques τ_{rij} of the motors. Using the development from [6], the rotor dynamics are given by $I_{rz}\dot{\Omega}_{rij} = -b\Omega_{rij}^2 + \tau_{rij}$ where I_{rz} is the rotor moment of inertia about the rotor z-axis, Ω_{rij} is the angular speed of rotor j, and b is a coefficient of friction due to aerodynamic drag on the rotor.

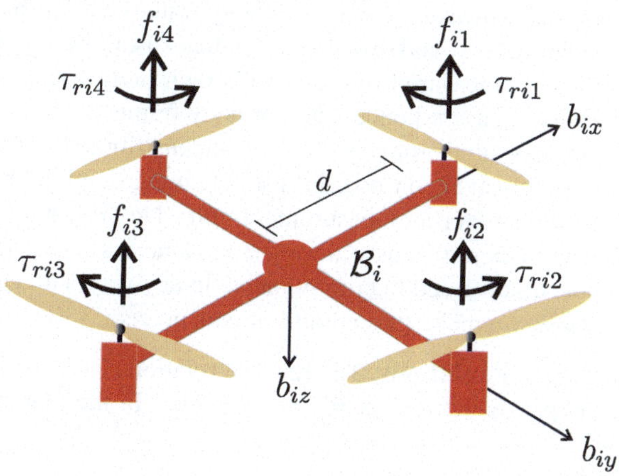

Fig. 2.5 Illustration of a quadrotor helicopter taken from [8] ©2012 AACC. Originally published in [1] and used here with permission of the American Automatic Control Council (AACC)

There is also an approximate algebraic relationship between the rotor thrust and rotor speed given by, $f_{ij} = \gamma \Omega_{rij}^2$, where γ is a parameter that can be experimentally determined. If we assume steady-state rotor dynamics such that $\dot{\Omega}_{rij} = 0$, then $f_{ij} = (\gamma/b)\tau_{rij} = c\tau_{rij}$ where $c = \gamma/b$ is the algebraic scaling factor between the rotor thrust and the applied motor torque. Using this fact, it is readily seen that the relationship between the control inputs and the motor torques is given by

$$\begin{bmatrix} u_i \\ \tau_{ix} \\ \tau_{iy} \\ \tau_{iz} \end{bmatrix} = \begin{bmatrix} c & c & c & c \\ 0 & -cd & 0 & cd \\ cd & 0 & -cd & 0 \\ 1 & -1 & 1 & -1 \end{bmatrix} \begin{bmatrix} \tau_{ri1} \\ \tau_{ri2} \\ \tau_{ri3} \\ \tau_{ri4} \end{bmatrix}.$$

In the above, the total thrust is equal to the summation of the four rotor thrusts; the torque about the b_{ix} axis is proportional to the differential thrust of the two rotors on the b_{iy} axis, $f_{i4} - f_{i2}$; the torque about the b_{iy} axis is proportional to the differential thrust of the two rotors on the b_{ix} axis, $f_{i1} - f_{i3}$; and the torque about the b_{iz} axis is equal to the summation of the four reaction torques which are equal and opposite to the applied motor torques τ_{rij}. With the definition of (u_i, τ_i) above, the quadrotor helicopter is modeled with (2.6), (2.7). △

2.3 Local and Distributed Feedback

Having presented the two robot models investigated in this monograph, we turn to the issue of sensing constraints. Since the robots we have in mind are to function autonomously and possibly on another planet, we seek to formulate coordination problems assuming minimal availability of sensors. In this monograph, a feedback controller meeting our sensing constraints is called *local and distributed.*

As we shall see in a moment, a local and distributed feedback for a kinematic unicycle is one that can only sense the unicycle's relative heading and relative displacement (measured in its own body frame) with respect to its neighbors. For flying vehicles, a local and distributed feedback is allowed to rely on relative positions and velocities (measured in the robot's own body frame) and the robot's own angular velocity, measured once again in body frame.

The essential property of a local and distributed feedback is that it can be computed using exclusively on-board sensors (cameras and, in the case of body frame angular velocity, an inertial measurement unit). No GPS units are assumed to be available, nor any external sensor network (a camera system, for example) is required. The robots do not need to communicate in any way with each other.

Now, we make these requirements precise:

- **Relative rotations.** The relative rotation (in $\mathbb{R}^2$ or $\mathbb{R}^3$) of robot j with respect to robot i is defined as $R_j^i := (R_i)^{-1} R_j$. In the planar case of rotation matrices in

SO(2), if we associate angles θ_i and θ_j in $\mathbb{S}^1$ to matrices R_i and R_j, then the angle associated with R^i_j is $\theta_j - \theta_i$.

- **Relative vectors.** If z_i is a vector (in $\mathbb{R}^2$ or $\mathbb{R}^3$) pertaining to robot i and z_j is another vector pertaining to robot j, both referenced in frame $\mathcal{I}$, then we define the relative vector z_{ij} as $z_{ij} := z_j - z_i$. This vector is referenced in the inertial frame $\mathcal{I}$. Recall that we denote by $z^i_{ij} := R_i^{-1} z_{ij}$ the coordinate representation of z_{ij} in frame $\mathcal{B}_i$.
  We will apply this notation to relative positions of unicycles and flying robots, x_{ij} and relative velocities of flying robots, v_{ij}.
- **Sensor graph.** Not every robot will be able to sense every other robot in the team. Instead, each robot can only sense a subset of neighbors. This sensing convention is naturally represented using the ideas from graph theory reviewed in Appendix B. We define the *sensor graph* $\mathcal{G} = (\mathcal{V}, \mathcal{E})$, where each node in the node set $\mathcal{V}$ represents a robot, and an edge in the edge set $\mathcal{E}$ between node i and node j indicates that robot i can sense robot j. We assume that $\mathcal{G}$ has no self-loops. Given a node i, its set of neighbors $\mathcal{N}_i$ represents the set of vehicles that robot i can sense.

Before we proceed with the definition of local and distributed feedback, we make an important remark about the nature of the sensor graph. In a realistic scenario, the neighbor set $\mathcal{N}_i$ would be the set of robots within the field of view of robot i. For instance, if each robot mounted an omnidirectional camera, then one could define $\mathcal{N}_i$ to be the collection of robots that are within a given distance from robot i. With such a definition, the sensor digraph $\mathcal{G}$ would be state-dependent, making the stability analysis of most coordination algorithms too hard at present. Relatively little research has been done on distributed coordination problems with state-dependent sensor graphs. In this context, in the simplest case when the robots are modeled as kinematic integrators, it has been shown in [9] that the circumcenter law of Ando et al. [10] preserves connectivity of the sensor graph and leads to rendezvous if the sensor graph is initially connected. Despite the simplicity of the robot model, the stability analysis in [9] is hard, and the control law is continuous but not Lipschitz continuous.

One could assume (as has been done in some literature) that the sensor graph is time varying, but this assumption would unnecessarily complicate the stability analysis without providing much insight, nor addressing the realistic scenario of a limited field of view. Since most of the coordination problems addressed in this book have remained unsolved until now, it is only natural to adopt the admittedly unrealistic assumption that the sensor graph is *static*, which will allow us to give rigorous stability assertions of various algorithms. We will then check in simulation to what extent one of our main coordination algorithms works well with state-dependent graphs and leave the theoretical investigation of state-dependent sensor graphs for future research.

We are now ready to define the notion of a local and distributed feedback.

Definition 2.1 A *local and distributed feedback* for the unicycle model of robot i in (2.5) is a locally Lipschitz function $(y^i_i, \varphi_i) \mapsto (u_i, \omega_i)$, where $y_i := (x_{ij})_{j\in\mathcal{N}_i}$, $y^i_i := (x^i_{ij})_{j\in\mathcal{N}_i}$ and $\varphi_i := (\theta_{ij})_{j\in\mathcal{N}_i}$. △

Thus, a local and distributed feedback for the unicycle is a function of relative displacements measured in body frame and relative headings.

Definition 2.2 A *local and distributed feedback* for flying robot i modeled as in (2.6)–(2.7) is a locally Lipschitz function $(y_i^i, \Omega_i^i) \mapsto (u_i, \tau_i)$, where $y_i := (x_{ij}, v_{ij})_{j \in \mathcal{N}_i}$ and $y_i^i := (x_{ij}^i, v_{ij}^i)_{j \in \mathcal{N}_i}$. △

A local and distributed feedback for the flying vehicle is a locally Lipschitz function of relative positions and velocities measured in body frame, as well as the vehicle's own angular velocity in body frame. The reader at this point might find it surprising that we do not include relative rotations R_j^i in the list of available sensor measurements for the flying robot. In principle, the inclusion of these quantities would still give a local and distributed feedback, but we would not need them for the algorithm we will present in Chap. 5.

A remark about the terminology "local and distributed." A *local feedback* is one in which all vectors are represented in the body frame of robot i, while a *distributed*

Table 2.1 Table of notation for kinematic unicycle model

Quantity	Description
$x_i \in \mathbb{R}^2$	Absolute position of unicycle i
$x \in \mathbb{R}^{2n}$	$(x_i)_{i \in \mathbf{n}}$
$R_i \in \mathsf{SO}(2)$	Attitude of unicycle i
$\theta_i \in \mathbb{S}^1$	Heading angle of unicycle i
$\theta \in \mathbb{T}^n$	$(\theta_i)_{i \in \mathbf{n}}$
$\omega_i \in \mathbb{R}$	Angular velocity of unicycle i
$y_i = (x_{ij})_{j \in \mathcal{N}_i}$	Vector of rel. pos. available to robot i
$\varphi_i = (\theta_{ij})_{j \in \mathcal{N}_i}$	Vector of rel. angles available to robot i

Table 2.2 Table of notation for flying robots.

Quantity	Description
$m_i,\ J_i$	Mass and inertia matrix of robot i
$x_i \in \mathbb{R}^3$	Inertial position of robot i
$x \in \mathbb{R}^{3n}$	$(x_i)_{i \in \mathbf{n}}$
$v_i \in \mathbb{R}^3$	Linear velocity of robot i
$v \in \mathbb{R}^{3n}$	$(v_i)_{i \in \mathbf{n}}$
$R_i \in \mathsf{SO}(3)$	Attitude of robot i
$\Omega_i \in \mathbb{R}^3$	Angular velocity of robot i
$\Omega \in \mathbb{R}^{3n}$	$(\Omega_i)_{i \in \mathbf{n}}$
$q_i = -R_i e_3$	Thrust direction vector of robot i
$T_i = -u_i R_i e_3$	Applied thrust vector of robot i
$y_i = (x_{ij}, v_{ij})_{j \in \mathcal{N}_i}$	Vector of rel. pos. and vel. available to robot i

feedback is one in which only relative quantities with respect to neighboring robots are accessible. In applications, a local and distributed feedback for robot i can be computed with on-board cameras. No information needs to be communicated between agents using a communication system or require centralized information.

Most results in this book will require local and distributed feedback, with some exceptions pointed out in Chap. 1. All the relevant quantities for the unicycle and flying robot models are summarized in Tables 2.1 and 2.2.

References

1. Roza, A., Maggiore, M.: Path following controller for a quadrotor helicopter. In: American Control Conference (ACC), pp. 4655–4660 (2012)
2. Roza, A., Maggiore, M., Scardovi, L.: Local and distributed rendezvous of underactuated rigid bodies. IEEE Trans. Autom. Control **62**(8), 3835–3847 (2017)
3. Hua, M., Hamel, T., Morin, P., Samso, C.: A control approach for thrust-propelled underactuated vehicles and its application to VTOL drones. IEEE Trans. Autom. Control **54**(8), 1837–1853 (2009)
4. Abdessameud, A., Tayebi, A.: Formation control of VTOL unmanned aerial vehicles with communication delays. Automatica **47**(11), 2383–2394 (2011)
5. Mokhtari, A., Benallegue, A., Orlov, Y.: Exact linearization and sliding mode observer for a quadrotor unmanned aerial vehicle. Int. J. Robot. Autom. **21**(1), 39–49 (2006)
6. Castillo, P., Lozano, R., Dzul, A.: Stabilization of a mini rotorcraft with four rotors. IEEE Control Syst. Mag. **25**(6), 45–55 (2005)
7. Abdessameud, A., Tayebi, A., Polushin, I.: Rigid body attitude synchronization with communication delays. In: 2012 American Control Conference (ACC), pp. 3736–3741 (2012)
8. Roza, A.: Motion control of rigid bodies in SE(3). Master's thesis, University of Toronto (2012)
9. Lin, Z., Francis, B., Maggiore, M.: State agreement for coupled nonlinear systems with time-varying interaction. SIAM J. Control Optim. **46**(1), 288–307 (2007)
10. Ando, H., Oasa, Y., Suzuki, I., Yamashita, M.: Distributed memoryless point convergence algorithm for mobile robots with limited visibility. IEEE Trans. Robot. Autom. **15**(5), 818–828 (1999)

Chapter 3
Coordination Problems

Abstract In this chapter we introduce the coordination problems that are the object of this monograph. For flying robots we introduce the rendezvous problem, and for unicycles we present problems in order of increasing complexity, starting with unicycle rendezvous followed by a number of formation control problems. A unifying feature of our treatment is that all coordination problems are cast in terms of stabilization of suitable subsets of the collective state space of the robots. Our formulation relies on stability notions presented in Appendix A, so the reader should consult this appendix before reading this chapter.

3.1 Rendezvous of Flying Robots

The objective of the rendezvous problem for flying robots is to design local and distributed feedbacks in the sense of Definition 2.2 making the positions of a team of flying robots converge to each other. We define the *rendezvous manifold* as

$$\begin{aligned}\Gamma := \Big\{ & (x_i, v_i, R_i, \Omega_i^i)_{i \in \mathbf{n}} \in \mathbb{R}^{3n} \times \mathbb{R}^{3n} \times \mathsf{SO}(3)^n \times \mathbb{R}^{3n} : x_{ij} = v_{ij} = 0, \\ & \Omega_i^i = \mathbf{\Omega}_i^i(y_i, R_i),\ i, j \in \mathbf{n} \Big\},\end{aligned} \tag{3.1}$$

where $\mathbf{\Omega}_i^i(y_i, R_i)$ is a suitable function to be designed in Chap. 5. There is no condition imposed on the relative attitudes.

Problem 1 (***R****endezvous* ***P****roblem for* ***F****lying Robots* (RP-F)) Consider a team of n flying robots modeled as in (2.6)–(2.7), and a static sensor digraph $\mathcal{G} = (\mathcal{V}, \mathcal{E})$ containing a globally reachable node. Design local and distributed feedbacks $(u_i^\star, \tau_i^\star) : (y_i^i, \Omega_i^i) \mapsto (u_i, \tau_i), i \in \mathbf{n}$ (see Table 2.1 for the definition of y_i), rendering Γ globally practically stable (GPS) under the constraint $(u_i^\star, \tau_i^\star)|_\Gamma = (0, 0)$. △

The goal of the rendezvous control problem is to achieve synchronization of the robot positions and velocities to any desired degree of accuracy from any initial condition. The requirement $(u_i^\star, \tau_i^\star)|_\Gamma = (0, 0)$ means that the robots are not actuated when rendezvous is achieved.

A. Roza et al., *Distributed Coordination Theory for Robot Teams*,
Lecture Notes in Control and Information Sciences 490,
https://doi.org/10.1007/978-3-030-96087-2_3

The remaining problems concern teams of kinematic unicycles. The table below gives a global picture of the problems (and associated abbreviations) that we are about to formulate.

Final motion \ Collective goal	Rendezvous	Full sync	Formation path following	Formation flocking
Rest	RP-U	PP		PP
Parallel		PPP, PFP, PFP-B	PPP	PFP, PFP-B
Circular		CPP, CFP	CPP	CFP

Abbreviation	Explanation	Target set
RP-U	**R**endezvous **P**roblem for Kinematic **U**nicycles	Γ
PP	**P**arallel Formation **P**roblem	Γ_{p}
PFP	**P**arallel Formation **F**locking **P**roblem	Γ_{pf}
PFP-B	**P**arallel Formation **F**locking **P**roblem with a **B**eacon	Γ_{pfb}
PPP	**P**arallel Formation **P**ath Following **P**roblem	Γ_{lp}
CFP	**C**ircular Formation **F**locking **P**roblem	Γ_{cf}
CPP	**C**ircular Formation **P**ath Following **P**roblem	Γ_{cp}

3.2 Rendezvous of Kinematic Unicycles

The rendezvous problem for kinematic unicycles is analogous to that for flying robots in the foregoing problem statement. We define the *rendezvous manifold*,

$$\Gamma := \left\{(x_i, \theta_i)_{i \in \mathbf{n}} \in \mathbb{R}^{2n} \times \mathbb{T}^n : x_{ij} = 0,\ i, j \in \mathbf{n}\right\}. \tag{3.2}$$

This is the subset of the collective state space of the unicycle robots in which all positions coincide with one another. As in the case of flying robots, there is no condition placed on the relative heading angles.

Problem 2 (***R**endezvous **P**roblem for Kinematic **U**nicycles* (RP-U)) Consider the team of kinematic unicycles in (2.4) or in (2.5), and a sensor digraph $\mathcal{G}$ containing a globally reachable node. Design local and distributed feedbacks $(u_i^\star, \omega_i^\star) : y_i^i \mapsto (u_i, \omega_i)$, $i \in \mathbf{n}$ (see Table 2.1 for the definition of y_i), rendering the rendezvous manifold Γ globally asymptotically stable (GAS) under the constraint $(u_i^\star, \omega_i^\star)|_\Gamma = (0, 0)$. △

The goal of the rendezvous problem is to drive the team of unicycles to some common location from any initial condition. The requirement $(u_i^\star, \omega_i^\star)|_\Gamma = (0, 0)$

in the rendezvous problem (RP-U) means that the unicycles stop when they rendezvous. This means that the unicycles do not consume any energy when rendezvous is achieved as one would expect from a good control strategy.

3.3 Formation Control Problems

In order to formulate various formation control problems, we need to describe what is meant by a formation. A *formation* of n unicycles is a geometric pattern defined modulo roto-translations by means of desired inter-agent displacements.

Let a fixed vector $d^1_{1i} \in \mathbb{R}^2$ denote the desired displacement of unicycle i relative to unicycle 1, measured in the frame $\mathcal{B}_1$ of unicycle 1, i.e., $d^1_{1i} := R_1^{-1}(x_i - x_1)$. We collect all the fixed relative displacements in a vector $d := (d^1_{1i})_{i \in \mathbf{2:n}} \in \mathbb{R}^{2(n-1)}$. As we argue below, the vector d specifies the formation modulo roto-translations, and so we refer to $d \in \mathbb{R}^{2(n-1)}$ as a *formation vector*.

To illustrate, consider the two formations of four unicycles illustrated in Fig. 3.1, labeled 1 and 2. These formations are related to one another through a rigid roto-translation in the inertial frame $\mathcal{I}$: to arrive at configuration 2, first rotate configuration 1 by angle φ about unicycle 1, followed by a translation of r units along the dotted line. For both configurations, the offsets $d^1_{1i} \in \mathbb{R}^2$ between unicycle 1 and unicycle i as measured in body frame 1, are identical, and therefore the two configurations represent the same formation. We correspondingly say that the formation is invariant under roto-translations.

The labeling of the unicycles is done solely for the purpose of defining the formation, and does not imply any attribution of priority to the unicycles. In some of the problems formulated below, it will be assumed, without loss of generality, that unicycle 1 is chosen to be at the front of the formation so that $d^1_{1i} \cdot e_1 \leq 0$ for all

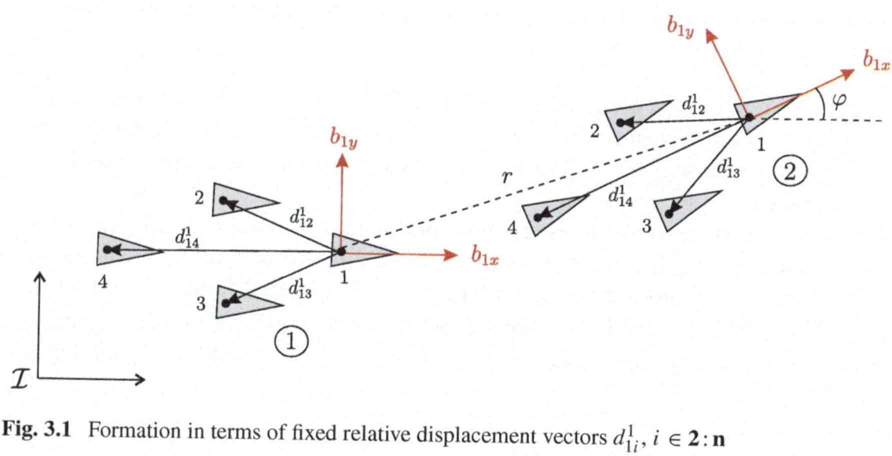

Fig. 3.1 Formation in terms of fixed relative displacement vectors d^1_{1i}, $i \in \mathbf{2:n}$

$i \in \mathbf{2}:\mathbf{n}$. This is the case in the example in Fig. 3.1. For a given formation vector d, we define the *formation manifold* as,

$$\Gamma := \left\{(x,\theta) \in \mathbb{R}^{2n} \times \mathbb{T}^n : x_{1i} = R_1 d^1_{1i},\ i \in \mathbf{n}\right\}. \tag{3.3}$$

The rendezvous manifold in (3.2) corresponds to the case that $d = 0$.

3.3.1 Parallel Formations That Stop

Let $d \in \mathbb{R}^{2(n-1)}$ be a formation vector, and without loss of generality, choose unicycle 1 to be at the front of the formation so that $d^1_{1i} \cdot e_1 \leq 0$ for all $i \in \mathbf{2}:\mathbf{n}$. Let

$$\mathsf{F} := \{d \in \mathbb{R}^{2(n-1)} : d^1_{1i} \cdot e_1 \leq 0,\ i \in \mathbf{2}:\mathbf{n}\}$$

be the corresponding set of all formation vectors. For any $d \in \mathsf{F}$, the objective of the formation control problem is to design local and distributed feedbacks to drive a team of unicycles to the *parallel formation manifold*,

$$\Gamma_p := \{(x,\theta) \in \Gamma : \theta_i = \theta_1,\ i \in \mathbf{2}:\mathbf{n}\}. \tag{3.4}$$

The parallel formation manifold Γ_p is the subset of the state space in which the unicycles have parallel headings, and their relative displacements meet the formation specification. A parallel formation is illustrated in Fig. 3.1.

Problem 3 (*Parallel Formation* ***P****roblem* (PP)) Consider the team of kinematic unicycles in (2.4) or in (2.5), and a connected, undirected sensor graph $\mathcal{G}$. For any choice of $d \in \mathsf{F}$, design local and distributed feedbacks $(u_i^\star, \omega_i^\star) : (y_i^i, \varphi_i) \mapsto (u_i, \omega_i), i \in \mathbf{n}$ (see Table 2.1 for the definition of y_i and φ_i), rendering the formation manifold Γ_p almost semiglobally asymptotically stable (ASGAS), under the constraint $(u_i^\star, \omega_i^\star)|_{\Gamma_p} = (0,0)$. △

As in the rendezvous problem, the requirement $(u_i^\star, \omega_i^\star)|_p = (0,0)$ means that the unicycles come to rest once the formation has been achieved. The problem of full synchronization in which both positions and headings synchronize for all unicycles is a special case of PP in which $d = 0$.

Now we introduce two formation control problems with parallel collective motion: parallel flocking and line path following. For flocking, the formation moves along a straight line whose roto-translation with respect to the inertial frame is not defined a priori. The specific line followed depends on the initial conditions of the agents. On the other hand, in path following, the formation follows a specific line path defined a priori in $\mathbb{R}^2$.

3.3.2 Parallel Formation Flocking

Let $d \in \mathsf{F}$ be a formation vector and $\bar{v} > 0$ be a desired flocking speed. The 2-tuple $(d, \bar{v})$ is a *parallel formation flocking specification*, and

$$\mathsf{PF} := \{(d, \bar{v}) \in \mathsf{F} \times \mathbb{R} : \bar{v} > 0\}$$

represents the set of all parallel flocking formations of n unicycles. For any $(d, \bar{v}) \in \mathsf{PF}$, the objective of the formation control problem is to design local and distributed feedbacks to drive a team of unicycles to a desired formation corresponding to the *parallel formation flocking manifold*

$$\Gamma_{pf} := \Gamma_p, \tag{3.5}$$

which coincides with the parallel formation manifold in Sect. 3.3.1. However, unlike PP where the formation is constrained to stop on the set Γ_p, for parallel formation flocking they are required to move at steady-state speed $\bar{v}$.

Problem 4 (*Parallel Formation **F**locking **P**roblem* (PFP)) Consider the team of kinematic unicycles in (2.4) or in (2.5), and sensor digraph $\mathcal{G}$ containing a globally reachable node. For any choice of $(d, \bar{v}) \in \mathsf{PF}$, design local and distributed feedbacks $(u_i^\star, \omega_i^\star) : (y_i^i, \varphi_i) \mapsto (u_i, \omega_i)$, $i \in \mathbf{n}$ (see Table 2.1 for the definition of y_i and φ_i), rendering Γ_{pf} almost globally asymptotically stable (AGAS), under the constraint $(u_i^\star, \omega_i^\star)|_{\Gamma_{pf}} = (\bar{v}, 0)$. △

As we shall see, we will only be able to solve PFP for a class of so-called hierarchical sensor graphs. Should a beacon be available, our solution will work for general digraphs. Accordingly, we next formulate a variation of PFP for the case when a beacon is available. More precisely, suppose each unicycle is permitted to measure, with respect to its own body frame, a common inertial vector p of unit norm, the beacon. That is, each unicycle can measure the vector $p^i = R_i^{-1} p$. The beacon p specifies a desired flocking direction. Let $\theta_p \in \mathbb{S}^1$ be the angle of p with respect to axis i_x of frame $\mathcal{I}$. The 3-tuple $(d, p, \bar{v})$, is a parallel formation flocking specification with a beacon, and

$$\mathsf{PFB} := \{(d, p, \bar{v}) \in \mathsf{F} \times \mathbb{R}^2 \times \mathbb{R} : \|p\| = 1,\ \bar{v} > 0\}$$

represents the set of all parallel flocking formations of n unicycles with a beacon. The set Γ_{pf} is replaced with the *parallel formation flocking manifold with a beacon*

$$\Gamma_{pfb} := \left\{(x, \theta) \in \Gamma : \theta_i = \theta_p,\ i \in \mathbf{n}\right\}, \tag{3.6}$$

in which not only do the unicycles have the same heading angle, therefore constituting a parallel formation, but the heading coincides with the angle of the beacon θ_p. The *Parallel Formation Flocking Problem with a Beacon* (PFP-B) is stated as follows.

Problem 5 (*Parallel Formation **F**locking **P**roblem with a* **B***eacon* (PFP-B)) Consider a team of n flying robots modeled as in (2.6)–(2.7), and sensor digraph $\mathcal{G}$ containing a globally reachable node. For any $(d, p, \bar{v}) \in \mathsf{PFB}$, design feedbacks $(u_i^\star, \omega_i^\star) : (y_i^i, \varphi_i, p^i) \mapsto (u_i, \omega_i)$, $i \in \mathbf{n}$ (see Table 2.1 for the definition of y_i and φ_i), rendering the set Γ_{pfb} AGAS under the constraint $(u_i^\star, \omega_i^\star)|_{\Gamma_{pfb}} = (\bar{v}, 0)$. △

When the formation vector d is chosen to be zero, both PFP and PFP-B reduce to full synchronization problems with nonzero asymptotic speed $\bar{v}$.

3.3.3 Parallel Formation Path Following

Let $d \in \mathsf{F}$ be a desired formation and $\mathcal{C}(r_0, p) = \{x \in \mathbb{R}^2 : x = r_0 + sp,\ s \in \mathbb{R}\} \subset \mathbb{R}^2$ be a line to be followed by one unicycle in the team (without loss of generality, unicycle 1) where $r_0, p \in \mathbb{R}^2$. Here p is a unit vector pointing tangent to $\mathcal{C}(r_0, p)$, and $\theta_p \in \mathbb{S}^1$ denotes the angle of p with respect to axis i_x of frame $\mathcal{I}$. Further, let $\bar{v} > 0$ be a desired path following speed. Assume that unicycle 1 is at the front of the formation so that $d_{1i}^1 \cdot e_1 \leq 0$ for all $i \in \mathbf{2}:\mathbf{n}$. The 4-tuple $(d, r_0, p, \bar{v})$ is a *formation line path following specification*, and

$$\mathsf{LP} := \{(d, r_0, p, \bar{v}) \in \mathsf{F} \times \mathbb{R}^2 \times \mathbb{R}^2 \times \mathbb{R} : \|p\| = 1,\ \bar{v} > 0\}$$

represents the set of all line following formations of n unicycles. We assume that unicycle i can measure the displacement vector between its position, x_i, and its orthogonal projection onto $\mathcal{C}(r_0, p)$, $c^\star(x_i)$, measured in body frame $\mathcal{B}_i$. This quantity is given by

$$\pi^i(x_i) := (c^\star(x_i) - x_i)^i = (r_0 - x_i)^i - ((r_0 - x_i)^i \cdot p^i)p^i. \tag{3.7}$$

For any $(d, r_0, p, \bar{v}) \in \mathsf{LP}$, we define the *formation line path following manifold*

$$\Gamma_{lp} := \left\{(x, \theta) \in \Gamma_{pfb} : x_1 \in \mathcal{C}(r_0, p)\right\}, \tag{3.8}$$

which coincides with Γ_{pfb} with the additional requirement that unicycle 1 lies on the path $\mathcal{C}$. An example of a configuration of unicycles in Γ_{lp} is illustrated in Fig. 3.2.

Problem 6 (*Parallel Formation **P**ath Following **P**roblem* (PPP)) Consider the team of kinematic unicycles in (2.4) or in (2.5), and sensor digraph $\mathcal{G}$ containing a globally reachable node. For any $(d, r_0, p, \bar{v}) \in \mathsf{LP}$, design feedbacks $(u_i^\star, \omega_i^\star) : (y_i^i, \varphi_i, p^i, \pi^i(x_i)) \mapsto (u_i, \omega_i)$, $i \in \mathbf{n}$ (see Table 2.1 for the definition of y_i and φ_i), rendering Γ_{lp} almost globally asymptotically stable (AGAS) under the constraint $(u_i^\star, \omega_i^\star)|_{\Gamma_{lp}} = (\bar{v}, 0)$. △

In the special case when the formation vector d is zero, PPP reduces to the full synchronization problem in which the unicycles converge to each other and move along the line $\mathcal{C}(r_0, p)$.

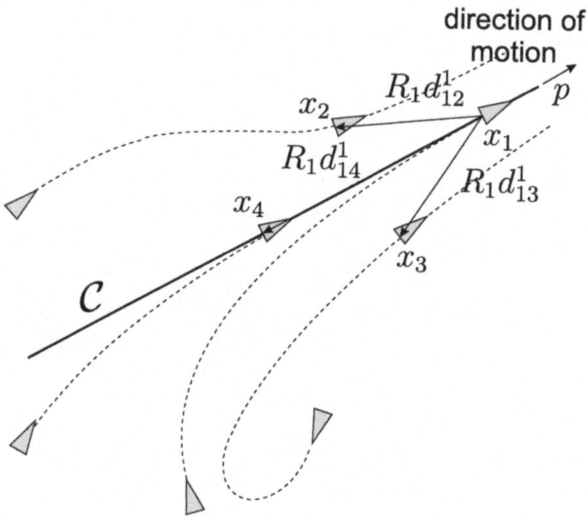

Fig. 3.2 Formation line path following

Next, we state two formation control problems with final circular collective motion: circular flocking and circular path following. For flocking, the formation moves around a circle of desired radius whose center is not defined a priori. Rather, it depends on the initial conditions of the agents. On the other hand, in circular path following, the formation follows a specific circle of desired radius whose center is defined a priori.

3.3.4 Circular Formation Flocking

The objective of the circular formation flocking control problem is to design local and distributed feedbacks (u_i, ω_i), $i \in \mathbf{n}$, making a team of unicycles converge to a desired formation with formation vector $d \in \mathbb{R}^{2(n-1)}$, and encircle, with angular speed $\bar{v} > 0$, a center point $c \in \mathbb{R}^2$ not defined a priori, and dependent on the initial conditions. For one unicycle in the team (without loss of generality, unicycle 1) we specify, a priori, the radius of rotation β_1 (see Fig. 3.3). The 3-tuple $(d, \beta_1, \bar{v})$ is a *circular formation flocking specification*, and

$$\mathsf{CF} := \{(d, \beta_1, \bar{v}) \in \mathbb{R}^{2(n-1)} \times \mathbb{R} \times \mathbb{R} : \beta_1 > 0,\ \bar{v} > 0\}$$

represents the set of all circular flocking formations of n unicycles.

When the desired formation is achieved, each unicycle $i \in \mathbf{2}:\mathbf{n}$ must move along a concentric circular orbit of radius $\beta_i = \sqrt{(d_{1i}^1 \cdot e_1)^2 + (\beta_1 - d_{1i}^1 \cdot e_2)^2}$ with angular speed $\bar{v}$. The radii β_i, $i \in \mathbf{n}$, are illustrated in Fig. 3.3. Unlike in the case of parallel flocking, where all unicycle headings converge to a common direction, in the

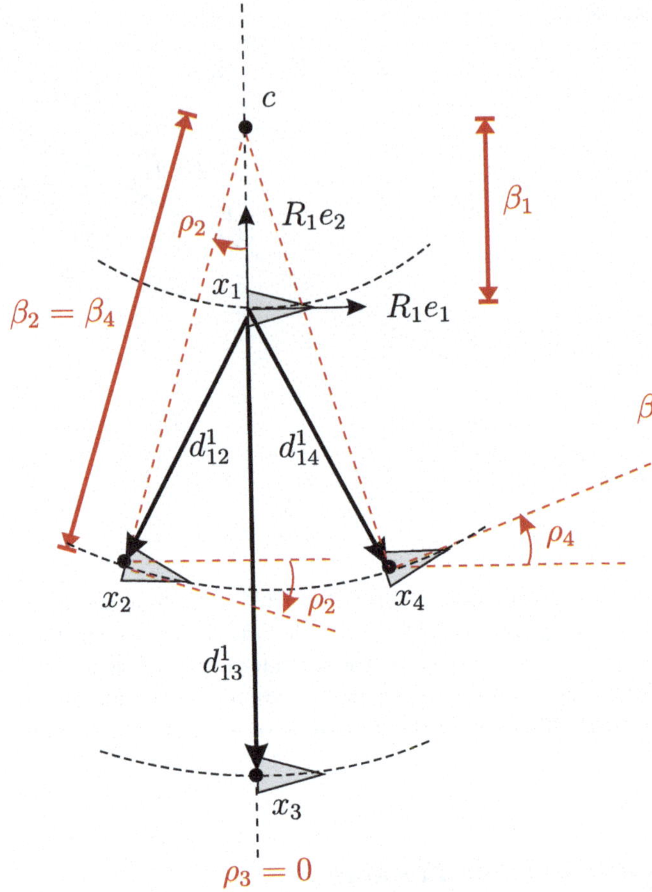

Fig. 3.3 Circular formation flocking with center of rotation c

case of circular formation flocking, the heading directions will not align in general. Instead, the heading of unicycle i is offset relative to the heading of unicycle 1 by the angle $\rho_i(d, \beta_1) = \text{atan2}(d^1_{1i} \cdot e_1, \beta_1 - d^1_{1i} \cdot e_2)$ shown in Fig. 3.3, where $\rho_2 = -\rho_4$ and $\rho_1 = \rho_3 = 0$.

Inspired by [1], define the offset vector $\delta_i = \beta_i R_i e_2$ for each unicycle i rigidly attached to its body frame and perpendicular to the heading direction $R_i e_1$. The vector $\hat{x}_i = x_i + \delta_i$ represents the perceived center of a circle of radius β_i that unicycle i would trace out if its translational speed were $\bar{v}\beta_i$, and its angular speed were $\bar{v}$. Define $\hat{x}_{ij} = \hat{x}_j - \hat{x}_i$. On the set $\Lambda := \{(x, \theta) \in \mathbb{R}^{2n} \times \mathbb{T}^n : \hat{x}_{1i} = 0,\ i \in \mathbf{n}\}$, unicycles lie on concentric circles of desired radii $(\beta_i)_{i \in \mathbf{n}}$ about a common center. Relative to this set, the desired formation is achieved if, in addition, $\theta_{1i} = \rho_i(d, \beta_1)$ for all $i \in \mathbf{2}:\mathbf{n}$. The *circular formation flocking manifold* is therefore given by,

$$\Gamma_{cf} := \{(x, \theta) \in \Lambda : \theta_{1i} = \rho_i(d, \beta_1),\ i \in \mathbf{2}:\mathbf{n}\}. \tag{3.9}$$

Problem 7 (*Circular Formation* ***F****locking* ***P****roblem* (CFP)) Consider the team of kinematic unicycles in (2.4) or in (2.5), and a connected, undirected sensor graph $\mathcal{G}$. For any $(d, \beta_1, \bar{v}) \in$ CF, design local and distributed feedbacks $(u_i^\star, \omega_i^\star) : (y_i^i, \varphi_i) \mapsto (u_i, \omega_i)$, $i \in \mathbf{n}$ (see Table 2.1 for the definition of y_i and φ_i), rendering Γ_{cf} AGAS under the constraint $(u_i^\star, \omega_i^\star)|_{\Gamma_{cf}} = (\bar{v}\beta_i, \bar{v})$. △

In the special case when the formation vector d is zero, CFP reduces to the full synchronization problem in which the unicycles converge to each other and move along some circle of radius β_1 asymptotically.

3.3.5 Circular Formation Path Following

For the circular formation following problem, the desired formation $d \in \mathbb{R}^{2(n-1)}$ follows a particular circular orbit with center $c \in \mathbb{R}^2$ defined a priori, as opposed to the unspecified center of the flocking problem. The 4-tuple $(d, \beta_1, c, \bar{v})$ is a *formation circle path following specification*, and

$$\mathsf{CP} := \{(d, \beta_1, c, \bar{v}) \in \mathbb{R}^{2(n-1)} \times \mathbb{R} \times \mathbb{R}^2 \times \mathbb{R} : \beta_1 > 0,\ \bar{v} > 0\}$$

represents the set of all circle following formations of n unicycles. The *formation circle path following manifold* is defined as

$$\Gamma_{cp} := \left\{(x, \theta) \in \Gamma_{cf} : \hat{x}_1 = c\right\}, \tag{3.10}$$

where the only difference from Γ_{cf} is that the perceived centers $\hat{x}_i = x_i + \delta_i$ not only coincide, but they coincide at the desired circle center c.

Problem 8 (*Circular Formation* ***P****ath Following* ***P****roblem* (CPP)) Consider the team of kinematic unicycles in (2.4) or in (2.5), and a connected, undirected sensor graph $\mathcal{G}$. For any $(d, \beta_1, c, \bar{v}) \in$ CP, design feedbacks $(u_i^\star, \omega_i^\star) : (y_i^i, \varphi_i, (c - x_i)^i) \mapsto (u_i, \omega_i)$, $i \in \mathbf{n}$ (see Table 2.1 for the definition of y_i), rendering the formation circle path following manifold Γ_{cp} almost globally asymptotically stable (AGAS) under the constraint $(u_i^\star, \omega_i^\star)|_{\Gamma_{cp}} = (\bar{v}\beta_i, \bar{v})$. △

Note that the class of feedbacks in CPP requires the measurement of the relative displacement of the unicycle to the center of the circle that needs to be followed, measured in body frame. In principle, this measurement can be obtained through onboard cameras.

In the special case when the formation vector d is zero, CPP reduces to the full synchronization problem in which the unicycles converge to each other and move along the circle of radius β_1 and center c.

Reference

1. El-Hawwary, M., Maggiore, M.: Distributed circular formation stabilization for dynamic unicycles. IEEE Trans. Autom. Control **58**(1), 149–162 (2013)

Chapter 4
Control Primitives

Abstract In the chapters that follow we will make use of elementary control functions to construct our control solutions for the problems introduced in Chap. 3. These control functions will be referred to as *control primitives* and will be the subject of this chapter. In particular, we will review consensus controllers for single integrators in $\mathbb{R}^2$ and rotational integrators in $\mathsf{SO}(2)$ as they are required to develop control solutions for kinematic unicycles. The control solutions for flying robots will instead make use of consensus controllers for double integrators in $\mathbb{R}^3$ and rotating rigid bodies in $\mathsf{SO}(3)$.

Table 4.1 summarizes the control primitives presented in this chapter. Constructing feedbacks out of simpler control primitives is a central theme in this book that will be seen time and again. This chapter relies on notions of graph theory presented in Appendix B, so the reader should consult this appendix before reading this chapter.

Throughout this chapter, the sensing convention is defined by a sensor graph $\mathcal{G} = (\mathcal{V}, \mathcal{E})$ introduced in Appendix B. Each node in the set $\mathcal{V}$ represents an agent, and an edge (i, j) in the edge set $\mathcal{E}$ between node i and node j indicates that agent i can sense agent j. That is, agent j lies in the neighbor set $\mathcal{N}_i$ of agent i.

4.1 Control Primitives for Single Integrators

4.1.1 Single Integrator Consensus Controllers

Consider n single integrators

$$\dot{x}_i = v_i, \ i \in \mathbf{n}, \tag{4.1}$$

where $x_i \in \mathbb{R}^2$ and $v_i \in \mathbb{R}^2$ is the control input of subsystem i. Let $x_{ij} := x_j - x_i$ and $x = (x_i)_{i \in \mathbf{n}} \in \mathbb{R}^{2n}$.

A. Roza et al., *Distributed Coordination Theory for Robot Teams*, Lecture Notes in Control and Information Sciences 490, https://doi.org/10.1007/978-3-030-96087-2_4

Table 4.1 Table of control primitives

Unicycle	
Single integrator consensus	(4.2)
Single integrator consensus—uniformly bounded control	(4.3)
Single integrator path following	(4.6)
Rotational integrator equilibrium stabilization	(4.12)
Rotational integrator consensus—Kuramoto	(4.14)
Rotational integrator consensus—AGAS	(4.15)
Flying robot	
Double integrator consensus	(4.10)
Rotating body equilibrium stabilization	(4.17)

Definition 4.1 The feedback

$$v_i = f_i((x_{ij})_{j \in \mathcal{N}_i}) := \sum_{j \in \mathcal{N}_i} a_{ij} x_{ij}, \tag{4.2}$$

where $a_{ij} > 0$ for all $j \in \mathcal{N}_i$, is an *integrator consensus controller* for (4.1) if $\mathcal{G}$ contains a globally reachable node.

The previous definition is motivated by a well-known result, stating that the consensus set

$$\Gamma_c := \{x \in \mathbb{R}^{2n} : x_i = x_j,\ i, j \in \mathbf{n}\}$$

is globally asymptotically stable for system (4.1), (4.2) if and only if $\mathcal{G}$ contains a globally reachable node [1–3].

The next definition introduces a different type of consensus controller that will be used in Chap. 7.

Definition 4.2 The feedback

$$v_i = f_i((x_{ij})_{j \in \mathcal{N}_i}) := \sum_{j \in \mathcal{N}_i} a_{ij} \frac{f(\|x_{ij}\|)}{\|x_{ij}\|} x_{ij}, \tag{4.3}$$

is a *uniformly bounded integrator consensus controller* if $a_{ij} = a_{ji} > 0$ and $f : \mathbb{R} \to \mathbb{R}$, the *interaction function*, is a locally Lipschitz function satisfying:

A1: $sf(s) > 0$ for all $s \neq 0$, $f(0) = 0$, and there exist $c_1, c_2 > 0$ such that $|f(s)| > c_1$ for all $|s| > c_2$.

A2: $\sup |f(s)| < \infty$.

Notice that the conditions on c_1 and c_2 in A1 imply that $f(s)$ cannot tend to zero as s tends to infinity. Each element $(x_{ij}/\|x_{ij}\|) f(\|x_{ij}\|)$ of the sum in (4.3) is continuous at $x_{ij} = 0$ because $f(s)$ is a continuous function and $f(0) = 0$ by assumption A1. We will omit the simple proof of the fact that each $f_i(\cdot)$ is Lipschitz continuous.

Examples of suitable interaction functions are $f(s) = \tanh(s)$ and

$$f(s) = \begin{cases} s, & \text{if } |s| \le 1 \\ s/|s| & \text{if } |s| > 1. \end{cases} \tag{4.4}$$

The following proposition shows that feedback (4.3) globally asymptotically stabilizes the consensus set Γ_c for any connected, undirected sensor graph $\mathcal{G}$.

Proposition 4.1 *Consider system* (4.1) *with a uniformly bounded integrator consensus controller* (4.3). *If the undirected sensor graph $\mathcal{G}$ is connected then the consensus set Γ_c is globally asymptotically stable.*

Proof Consider system (4.1) with feedback (4.3). The feedback f_i in (4.3) for unicycle i points into the convex hull formed by its neighbors. By Corollary 3.9 in [4], the group of unicycles for system (4.1) achieves global consensus. □

4.1.2 Path Following Controllers

Consider the single integrator,

$$\dot{x} = v, \tag{4.5}$$

where $x \in \mathbb{R}^2$ and $v \in \mathbb{R}^2$ is the control input. Define a smooth simple (no self-intersections) curve $\mathcal{C} \subset \mathbb{R}^2$ and a final speed $\bar{v} > 0$.[1]

Definition 4.3 The feedback

$$v = h(x), \tag{4.6}$$

where $h : \mathbb{R}^2 \to \mathbb{R}^2$, is an *integrator path following controller* for $\mathcal{C}$ if it satisfies the following properties:

A1: h is globally Lipschitz; i.e., there exists $c > 0$ such that for all $x_1, x_2 \in \mathbb{R}^2$, $\|h(x_1) - h(x_2)\| \le c\|x_1 - x_2\|$,
A2: $h(x) = \bar{v}\vec{r}(x)$ for all $x \in \mathcal{C}$,
A3: the set $\mathcal{C}$ is asymptotically stable for (4.5), (4.6).

In this book, we will consider paths that are straight lines of the form $\mathcal{C}(r_0, p) = \{x \in \mathbb{R}^2 : x = r_0 + sp,\ s \in \mathbb{R}\} \subset \mathbb{R}^2$ for $r_0, p \in \mathbb{R}^2$. Single integrators are essentially the simplest model that one can design a path following controller for.

For a line $\mathcal{C}(r_0, p)$, we can define an *integrator line following controller*

$$h(x) = k_0(r_0 - x) - k_0((r_0 - x) \cdot p)p + \bar{v}p \tag{4.7}$$

[1] If $x = \sigma(s)$ is an arc length parametrization of $\mathcal{C}$, then we define the unit vector tangent to $\mathcal{C}$ by $\vec{r}(x) = \vec{r}(\sigma(s)) = (d/ds)\sigma(s)$.

with $k_0 > 0$, which can be rewritten as

$$h(x) = k_0(c^\star(x) - x) + \bar{v}p, \tag{4.8}$$

where $c^\star(x)$ is the orthogonal projection of the point x onto $\mathcal{C}(r_0, p)$, i.e., $(c^\star(x) - x) \cdot p = 0$. It is easy to verify that h in (4.7) satisfies properties A1–A3. Therefore, controller (4.8) makes x converge to the path $\mathcal{C}(r_0, p)$ moving in the direction of the beacon p with steady-state speed $\bar{v}$.

4.2 Control Primitives for Double Integrators

4.2.1 Double Integrator Consensus

Consider a collection of n double integrators

$$\begin{aligned} \dot{x}_i &= v_i, \\ \dot{v}_i &= u_i, \;\; i \in \mathbf{n}, \end{aligned} \tag{4.9}$$

where $x_i \in \mathbb{R}^3$, $v_i \in \mathbb{R}^3$ and $u_i \in \mathbb{R}^3$ is the control input of subsystem i. Let $x_{ij} := x_j - x_i$, $v_{ij} := v_j - v_i$, $x := (x_i)_{i \in \mathbf{n}} \in \mathbb{R}^{3n}$ and $v := (v_i)_{i \in \mathbf{n}} \in \mathbb{R}^{3n}$.

The feedback

$$u_i = f_i((x_{ij}, v_{ij})_{j \in \mathcal{N}_i}) := \sum_{j \in \mathcal{N}_i} a_{ij}\Big(x_{ij} + \gamma v_{ij}\Big), \;\; i \in \mathbf{n}, \tag{4.10}$$

where $a_{ij}, \gamma > 0$ is a *double integrator consensus controller*. The following theorem is taken from Ren et al. in [5, Theorems 4.1, 4.2] and says that for sufficiently large γ, if the sensing digraph contains a globally reachable node, then the system of double integrators achieves consensus.

Theorem 4.1 ([5] *Consider system* (4.9) *with feedback* (4.10) *and sensor digraph* $\mathcal{G}$ *containing a globally reachable node with corresponding weighted Laplacian matrix* L *with weights* $a_{ij} > 0$. *Suppose that* γ *is chosen to satisfy*

$$\gamma > \max_{\mu_i \neq 0} \sqrt{\frac{2}{|\mu_i| \cos\left(\frac{\pi}{2} - \tan^{-1} \frac{-\mathfrak{Re}(\mu_i)}{\mathfrak{Im}(\mu_i)}\right)}},$$

where $(\mu_i)_{i\in\mathbf{n}}$ are the eigenvalues of $-L$. Then the set

$$\left\{(x, v) \in \mathbb{R}^{3n} \times \mathbb{R}^{3n} : x_i = x_j, v_i = v_j,\ i, j \in \mathbf{n}\right\}$$

is globally asymptotically stable.

4.3 Control Primitives for Rotational Integrators

While single and double integrators have a state-space which is a non-compact manifold, rotational integrators and rotating bodies evolve on compact manifolds without boundary. This fact leads to a topological obstruction in which global asymptotic consensus or equilibrium stabilization cannot be achieved using continuous, time-invariant control [6]. In this case, almost global asymptotic stability is the best one can do with continuous time-invariant control. As a result, one may expect that a control solution built out of a rotational integrator control primitive might inherit this characteristic.

4.3.1 Rotational Integrator Equilibrium Stabilization

Consider a rotational integrator

$$\dot{\theta} = \omega, \tag{4.11}$$

where $\theta \in \mathbb{S}^1$ and ω is the control input. A feedback

$$\omega = g(\theta) := -k_0 \sin(\theta), \tag{4.12}$$

where $k_0 > 0$, is a *rotational integrator equilibrium stabilizer*. The following proposition states that the equilibrium $\theta = 0$ is almost globally asymptotically stable.

Proposition 4.2 *Consider system* (4.11) *with feedback* (4.12). *The union of the two equilibria $\theta = 0$ and $\theta = \pi$ is globally attractive. The first equilibrium is almost globally asymptotically stable while the second is exponentially unstable.*

Proof The result follows by a standard Lyapunov analysis using the Lyapunov function $V = 1 - \cos(\theta)$. The equilibrium $\theta = \pi$ is exponentially unstable since computing the linearization at $\theta = \pi$ gives

$$\frac{\mathrm{d}}{\mathrm{d}\theta}(-k_0 \sin\theta)|_{\theta=\pi} = -k_0 \cos\theta|_{\theta=\pi} = -k_0 \cos\pi = k_0 > 0.$$

□

4.3.2 Rotational Integrator Consensus

Consider a collection of rotational integrators,

$$\dot{\theta}_i = \omega_i, \; i \in \mathbf{n}, \tag{4.13}$$

where $\theta_i \in \mathbb{S}^1$, $\theta := (\theta_i)_{i\in\mathbf{n}} \in \mathbb{T}^n$, $\theta_{ij} := \theta_j - \theta_i$ and ω_i is the control input of subsystem i.

Definition 4.4 A feedback

$$\omega_i = g_i((\theta_{ij})_{j\in\mathcal{N}_i}) := \sum_{j\in\mathcal{N}_i} b_{ij} \sin(\theta_{ij}) \tag{4.14}$$

with $b_{ij} = b_{ji} > 0$ is a *Kuramoto consensus controller.*

System (4.13) with (4.14) is the well-known Kuramoto model [7] where the natural frequencies of the oscillators are set to zero.

Proposition 4.3 *([4]) Consider system* (4.13) *with feedback* (4.14) *and connected, undirected sensor graph* $\mathcal{G}$*. The set* $\{\theta \in \mathbb{T}^n : \theta_i = \theta_j, \; i, j \in \mathbf{n}\}$ *is asymptotically stable with domain of attraction containing*

$$S_\pi := \left\{\theta \in \mathbb{T}^n : |\theta_{ij}| < \pi, \; i, j \in \mathbf{n}\right\},$$

the set where all vehicle angles lie in the same half plane on an open arc of π *radians.*

Proof Without loss of generality, assume $\theta_i \in (-\pi/2, \pi/2) \subset \mathbb{S}^1$ for all $i \in \mathbf{n}$ which is diffeomorphic to the open segment $(-\pi/2, \pi/2) \in \mathbb{R}$. If the angle of a unicycle i is greater than that of all its neighbors, then g_i will be negative and its angle will decrease. The opposite is true for a unicycle whose angle is less than all its neighbors. Therefore, for all unicycles $i \in \mathbf{n}$, the input g_i points into the convex hull formed by its neighbors. By [4], the group of unicycles for system (4.13) achieves consensus with the feedback in (4.14) on $(-\pi/2, \pi/2)$. □

Unfortunately, the Kuramoto consensus controller of Definition 4.4 is not sufficient to construct the control solutions presented in Chap. 7, because in that chapter we require almost global stabilizers of the set $\{\theta \in \mathbb{T}^n : \theta_i = \theta_j, \; i, j \in \mathbf{n}\}$. Therefore, with the next definition, we introduce a different class of consensus controllers on $\mathbb{T}^n$.

Definition 4.5 The feedback

$$\omega_i = g_i((\theta_{ij})_{j\in\mathcal{N}_i}, \eta) := \eta_i \sum_{j\in\mathcal{N}_i} b_{ij} g(\theta_{ij}) \tag{4.15}$$

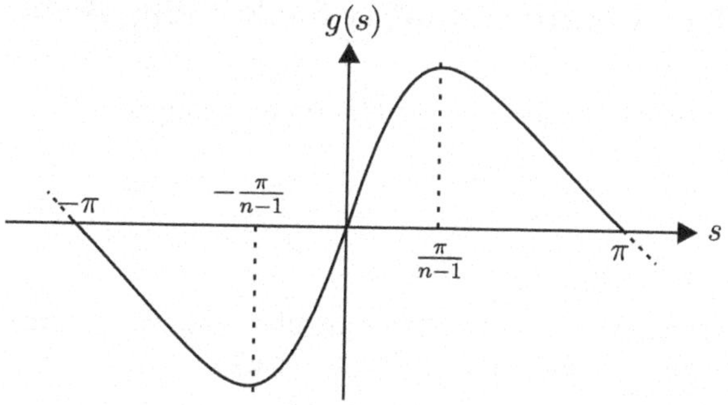

Fig. 4.1 Illustration of properties B1, B2 and B3. ©2018 IEEE. Reprinted, with permission, from [9]

is a *rotational integrator consensus controller* if (i) $\eta_i > 0$ where $\eta := (\eta_i)_{i \in \mathbf{n}}$, (ii) $b_{ij} = b_{ji} > 0$, and (iii) $g : \mathbb{S}^1 \to \mathbb{R}$ is a continuously differentiable *interaction function* satisfying the following three assumptions [8]:

B1: $sg(s) > 0$ for all $s \in (-\pi, \pi)\backslash 0$, $g(0) = g(\pi) = 0$;
B2: $g(s)$ is an odd function: $g(-s) = -g(s)$ for $s \in (-\pi, \pi)$;
B3: $\dot{g}(s) > 0$, $\forall s \in (-\frac{\pi}{n-1}, \frac{\pi}{n-1})$ and $\dot{g}(s) < 0$, $\forall s \in (-\pi, -\frac{\pi}{n-1}) \cup (\frac{\pi}{n-1}, \pi)$.

A sample interaction function satisfying B1-B3 is shown in Fig. 4.1.

The Kuramoto consensus controller corresponds to the choice $\eta_i = 1$ for all $i \in \mathbf{n}$ and $g(s) = \sin(s)$ and satisfies properties B1–B2, but does not satisfy property B3 for $n > 2$. In [8], Mallada–Freeman–Tang showed that a feedback enjoying properties B1–B3 almost globally stabilizes the set $\{\theta \in \mathbb{T}^n : \theta_i = \theta_j,\ i, j \in \mathbf{n}\}$ for almost all gains $b_{ij} = b_{ji} > 0$. The following result is a special case of Theorem 2 in [8].

Theorem 4.2 ([8, 10]) *Let $\mathcal{G}$ be an undirected and connected sensor graph and consider system* (4.13) *with feedback* (4.15) *satisfying assumptions B1–B3, expressed in states relative to agent one, i.e., $\tilde{\theta} := (\theta_{1i})_{i \in \mathbf{2:n}}$,*

$$\dot{\theta}_{1i} = g_i((\theta_{ij})_{j \in \mathcal{N}_i}, \eta) - g_1((\theta_{1j})_{j \in \mathcal{N}_1}, \eta),\ i \in \mathbf{2:n}.$$

There exists a set $N_b \subset (\mathbb{R}_+)^{|\mathcal{E}|}$ of Lebesgue measure zero such that for any collection of gains $(b_{ij})_{(i,j) \in \mathcal{E}} \in (\mathbb{R}_+)^{|\mathcal{E}|} \backslash N_b$, and for any $\eta_i > 0$, $i \in \mathbf{n}$, the origin $\tilde{\theta} = 0$ is almost globally asymptotically stable. Moreover, there exists a compact set of isolated equilibria $\mathcal{A}$ such that $\{0\} \cup \mathcal{A}$ is globally attractive and the equilibria in $\mathcal{A}$ are exponentially unstable.

The above result follows from Theorem 2 in [8]. In particular, system (4.13) with feedback (4.15) corresponds to the model in Eqs. (14)–(16) in [8] by letting $\chi_i(s) = \eta_i s$, letting ζ be the identity function and eliminating the integrator state γ_i.

The main difference here, compared to the solution in [8, Theorem 2], is that in [8] each system has an additional constant bias. The integrator state γ_i is used in [8] to compensate for this bias. In this work, system (4.13) has zero bias, so the integrator state γ_i is not needed. Accounting for this small difference, the proof of Theorem 4.2 follows from minimal modifications to the proof of [8, Theorem 2].

4.4 Control Primitive for Rotating Bodies in SO(3)

Consider the dynamic system for a rotating axis $\Gamma \in \mathbb{S}^2$ in three dimensions

$$\begin{aligned} \dot{\Gamma} &= \Gamma \times \omega, \\ J\dot{\omega} &= \tau - \omega \times J\omega, \end{aligned} \tag{4.16}$$

where $\omega \in \mathbb{R}^3$ is the body's angular velocity, τ is the torque control input and J is a symmetric inertia matrix. A feedback

$$\tau = \tau_d(\Gamma, \omega) \tag{4.17}$$

with smooth function $\tau_d : \mathbb{R}^3 \times \mathbb{R}^3 \to \mathbb{R}^3$ is a *rotating body equilibrium stabilizer* if the closed-loop system (4.16), (4.17) has an AGAS equilibrium point $(\Gamma, \omega) = (e_3, 0)$.

An example of a thrust direction controller is presented in [11] as,

$$\tau_d(\Gamma, \omega) = k_q\,(e_3 \times \Gamma) - K_\omega \omega, \tag{4.18}$$

where k_q is positive and K_ω is a symmetric and positive definite matrix.

Proposition 4.4 (Theorem 2 in *[11]*) *Feedback* (4.18) *is a rotating body equilibrium stabilizer.*

References

1. Ren, W., Beard, R.: Consensus seeking in multiagent systems under dynamically changing interaction topologies. IEEE Trans. Autom. Control **50**(5), 655–661 (2005)
2. Moreau, L.: Stability of continuous-time distributed consensus algorithms. In: Proceedings of the 43rd IEEE Conference on Decision and Control, pp. 3998–4003 (2004)
3. Olfati-Saber, R., Murray, R.: Consensus problems in networks of agents with switching topology and time-delays. IEEE Trans. Autom. Control **49**(9), 1520–1533 (2004)
4. Lin, Z., Francis, B., Maggiore, M.: State agreement for continuous-time coupled nonlinear systems. SIAM J. Control Optim. **46**(1), 288–307 (2007)
5. Ren, W., Atkins, E.: Distributed multi-vehicle coordinated control via local information exchange. Int. J. Robust Nonlinear Control **17**(10–11), 1002–1033 (2007)
6. Bhat, S., Bernstein, D.: A topological obstruction to continuous global stabilization of rotational motion and the unwinding phenomenon. Syst. Control Lett. **39**(1), 63–70 (2000)
7. Kuramoto, Y.: Chemical Oscillators, Waves, and Turbulence. Springer, Berlin (1984)
8. Mallada, E., Freeman, R., Tang, A.: Distributed synchronization of heterogeneous oscillators on networks with arbitrary topology. IEEE Trans. Control Netw. Syst. **3**(1), 12–23 (2016)

9. Roza, A., Maggiore, M., Scardovi, L.: A smooth distributed feedback for global rendezvous of unicycles. IEEE Trans. Control Netw. Syst. **5**(1), 640–652 (2018)
10. Mallada, E., Tang, A.: Synchronization of weakly coupled oscillators: coupling, delay and topology. J. Phys. Math. Theor. **46**(50) (2013)
11. Chaturvedi, N., Sanyal, A., McClamroch, N.: Rigid-body attitude control. IEEE Control Syst. Mag. **31**, 30–51 (2011)

Chapter 5
Rendezvous of Flying Robots

Abstract In this chapter, we present a solution to the rendezvous control problem for flying robots (RP-F), in which the objective is to make a group of underactuated flying robots convene at a common location by means of a local and distributed feedback. The solution we present is the first for this kind of problem, and it makes the rendezvous manifold globally practically stable.

5.1 Review of the Rendezvous Control Problem

Recall the model of a team of flying robots developed in Chap. 2,

$$\begin{aligned} \dot{x}_i &= v_i, \\ m_i \dot{v}_i &= -u_i R_i e_3 + m_i g, \end{aligned} \tag{5.1}$$

$$\begin{aligned} \dot{R}_i &= R_i\, (\Omega_i^i)^{\times}, \\ J_i \dot{\Omega}_i^i &= \tau_i - \Omega_i^i \times J_i \Omega_i^i, \quad i \in \mathbf{n}. \end{aligned} \tag{5.2}$$

The schematic representation of one such robot is in Fig. 5.1.

The goal of the rendezvous problem (RP-F), defined in Sect. 3.1, is to design local and distributed feedbacks making the positions of the robots in the team converge to each other. The rendezvous manifold is defined as

$$\begin{aligned} \Gamma := \Big\{ &(x_i, v_i, R_i, \Omega_i^i)_{i\in\mathbf{n}} \in \mathbb{R}^{3n} \times \mathbb{R}^{3n} \times \mathsf{SO}(3)^n \times \mathbb{R}^{3n} : x_{ij} = v_{ij} = 0, \\ &\Omega_i^i = \boldsymbol{\Omega}_i^i(y_i, R_i),\ i, j \in \mathbf{n} \Big\}, \end{aligned} \tag{5.3}$$

where $\boldsymbol{\Omega}_i^i(y_i, R_i)$ is a suitable function to be designed in the next section. In this

A. Roza et al., *Distributed Coordination Theory for Robot Teams*, Lecture Notes in Control and Information Sciences 490, https://doi.org/10.1007/978-3-030-96087-2_5

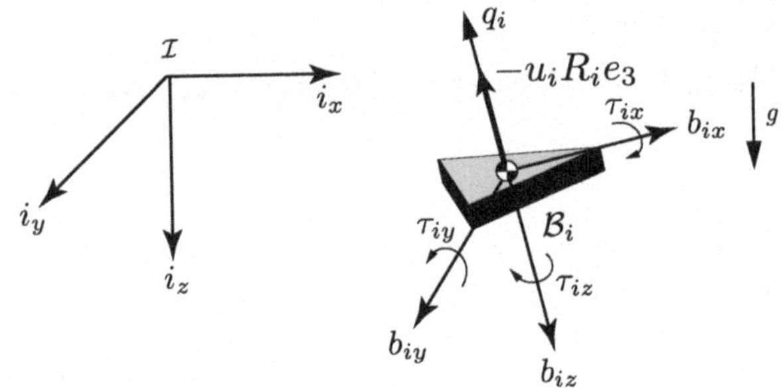

Fig. 5.1 Flying robot class. ©2017 IEEE. Reprinted, with permission, from [1]

section, we design a local and distributed feedback that renders Γ globally practically stable under the constraint that the robots' control inputs (u_i, τ_i) are zero when rendezvous is achieved.

5.2 Solution of the Rendezvous Control Problem

As outlined in the previous chapters, we make use of control primitives to derive the control solutions in this book. In particular, to solve RP-F we leverage the consensus controller for double integrators introduced in Chap. 4, which we repeat here for convenience,

$$f_i(y_i) = \sum_{j \in \mathcal{N}_i} a_{ij}\Big(x_{ij} + \gamma v_{ij}\Big), \; i \in \mathbf{n}. \tag{5.4}$$

We begin with the observation that the double integrator consensus controller (5.4) makes the system

$$\begin{aligned} \dot{x}_i &= v_i, \\ \dot{v}_i &= f_i + g, \; i \in \mathbf{n} \end{aligned} \tag{5.5}$$

achieve rendezvous, since the addition of the gravity vector g is common to all agents and does not affect the relative dynamics. Now compare system (5.5) to the translational dynamics of the flying robots,

$$\begin{aligned} \dot{x}_i &= v_i, \\ \dot{v}_i &= \frac{1}{m_i} u_i q_i + g, \; i \in \mathbf{n}, \end{aligned} \tag{5.6}$$

where $q_i = -R_i e_3$ is the thrust direction vector. Observe that if it were the case that $f_i = (1/m_i) u_i q_i$, systems (5.5) and (5.6) would be identical, and therefore, setting $u_i q_i = m_i f_i(y_i)$ in (5.6) would solve RP-F.

Following this intuition, the main idea behind our control solution is to drive $u_i q_i$ approximately to the desired thrust $m_i f_i(y_i)$. To this end, we pick the thrust magnitude

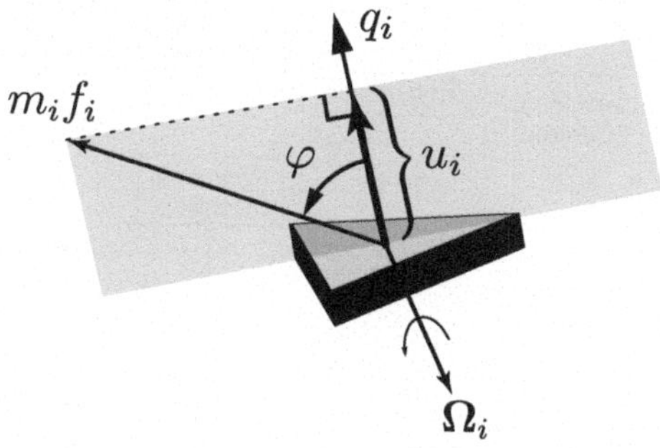

Fig. 5.2 Illustration of the control input u_i and reference angular velocity $\boldsymbol{\Omega}_i$ in (5.7). ©2017 IEEE. Reprinted, with permission, from [1]

u_i to be the projection of the desired thrust $m_i f_i(y_i)$ onto the thrust direction vector $u_i = m_i f_i(y_i) \cdot q_i$ (see Fig. 5.2), and we design the rotational control τ_i to align $u_i q_i$ with $m_i f_i(y_i)$.

Before outlining the design of the rotational control, we need to express the thrust magnitude u_i in body coordinates. Observe that the double integrator consensus controller (5.4) satisfies

$$f_i(y_i) = \sum_{j \in \mathcal{N}_i} a_{ij}(x_{ij} + \gamma v_{ij}) = \sum_{j \in \mathcal{N}_i} a_{ij}(R_i x^i_{ij} + \gamma R_i v^i_{ij}) = R_i f_i(y^i_i).$$

Using the fact that dot products are invariant under rotations, we have

$$u_i = m_i f_i(y_i) \cdot q_i = m_i(R_i f_i(y^i_i)) \cdot (-R_i e_3) = -m_i f_i(y^i_i) \cdot e_3 =: u^\star_i(y^i_i, \Omega^i_i),$$

which is the *thrust feedback* expressed in body coordinates.

We now proceed to present the intuition behind the design of the rotational control τ_i. Remember that the objective is to drive $u_i q_i$ to $m_i f_i(y_i)$. Define

$$\boldsymbol{\Omega}_i(y_i, R_i) := k_1(q_i \times f_i(y_i)),$$

which, in body coordinates, reads

$$\boldsymbol{\Omega}^i_i(y_i, R_i) = R_i^{-1} k_1 \left(f_i(y_i) \times R_i e_3\right) = k_1 \left(f_i(y^i_i) \times e_3\right).$$

Notice that the vector $\boldsymbol{\Omega}_i$ is perpendicular to the plane formed by the thrust direction vector q_i and the desired thrust force $m_i f_i(y_i)$, as shown in Fig. 5.2. Since the angular velocity vector identifies an instantaneous axis of rotation, it follows that if $\Omega_i = \boldsymbol{\Omega}_i$, then robot i rotates about $\boldsymbol{\Omega}_i$ according to the right-hand rule. Referring to Fig. 5.2, we see that such a rotation would close the gap between $u_i q_i$ and $m_i f_i(y_i)$, and the speed of rotation would be proportional to $\sin\varphi$, where φ is the angle between $u_i q_i$ and $m_i f_i(y_i)$ marked in the figure. When the gap is closed, we have $u_i = \|m_i f_i(y_i)\|$, $q_i = m_i f_i(y_i)/\|m_i f_i(y_i)\|$, and thus $u_i q_i = m_i f_i(y_i)$.

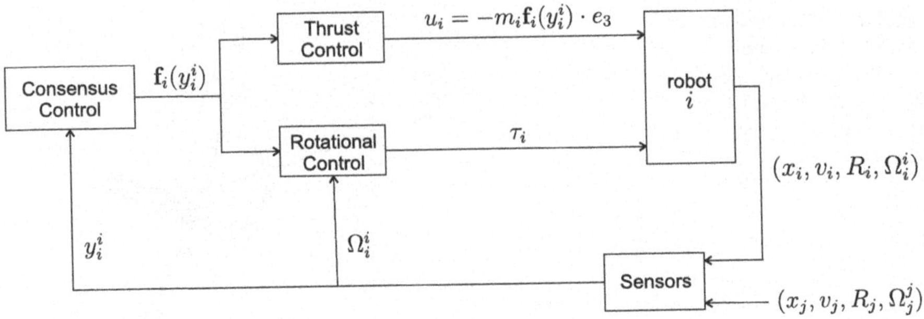

Fig. 5.3 Block diagram of the rendezvous control system for robot i. The outer loop assigns a desired thrust vector $f_i(y_i^i)$. The inner loop thrust control uses $f_i(y_i^i)$ to assign the vehicle input u_i while the rotational control uses $f_i(y_i^i)$ to assign the torque input τ_i. The vector y_i^i contains the relative displacements and velocities of vehicles that robot i can sense, measured in the body frame of robot i. ©2017 IEEE. Reprinted, with permission, from [1]

The last step is then to design the torque inputs τ_i to make Ω_i^i converge to an arbitrarily small neighborhood of $\boldsymbol{\Omega}_i^i$, $i \in \mathbf{n}$. It turns out that the following torque inputs achieve the desired objective

$$\begin{aligned}\tau_i =& \tau_i^\star(y_i^i, \Omega_i^i)\\ =& \Omega_i^i \times J_i \Omega_i^i - k_1 J_i \left((\Omega_i^i \times f_i(y_i^i)) \times e_3\right) - k_1^2 k_2 \left[\Omega_i^i - k_1(f_i(y_i^i) \times e_3)\right], \; i \in \mathbf{n}.\end{aligned}$$

To summarize, our proposed feedback is

$$\begin{aligned}u_i =& u_i^\star(y_i^i, \Omega_i^i) = -m_i f_i(y_i^i) \cdot e_3,\\ \tau_i =& \tau_i^\star(y_i^i, \Omega_i^i)\\ =& \Omega_i^i \times J_i \Omega_i^i - k_1 J_i \left((\Omega_i^i \times f_i(y_i^i)) \times e_3\right) - k_1^2 k_2 \left[\Omega_i^i - k_1(f_i(y_i^i) \times e_3)\right], \; i \in \mathbf{n}.\end{aligned} \tag{5.7}$$

The resulting control scheme leads to the modular solution depicted in Fig. 5.3. The outer loop of the block diagram assumes that $u_i q_i$ is the control input of (5.6) and computes a desired double integrator force $m_i f_i(y_i)$, which becomes a reference signal for an inner loop composed of the thrust and a rotational controllers.

The theorem below is the main result of this chapter, and it states that for sufficiently large $k_1, k_2 > 0$, the feedbacks in (5.7) solve RP-F if the network of robots has a sensor digraph containing a globally reachable node.

Theorem 5.1 RP-F *is solvable for system* (5.1), (5.2) *if and only if the sensor digraph $\mathcal{G}$ contains a globally reachable node, in which case a solution is the following. Let $f_i(y_i)$, $i \in \mathbf{n}$, be a double integrator consensus controller in* (5.4) *satisfying the conditions in Theorem 4.1. The local and distributed feedback in* (5.7) *where $k_1, k_2 > 0$ are control parameters, makes the rendezvous manifold (3.1) globally* *practically stable; that is, for any* $\varepsilon > 0$, *there exist positive gains* $k_1^\star > 0,\ k_2^\star > 0$ *such that for all* $k_1 > k_1^\star$, $k_2 > k_2^\star$, *the set* $B_\varepsilon(\Gamma)$ *has a globally asymptotically stable subset containing* Γ.

While the intuition behind the proposed controller is simple, the proof that the interplay between the two nested loops results in global practical stability of the rendezvous manifold is rather delicate, and it crucially relies on the homogeneity of the functions $f_i(y_i)$, $i \in \mathbf{n}$. Below we provide the reader with an outline of the proof of Theorem 5.1. For the full proof, we refer the reader to our paper [1].

5.3 Outline of the Proof of Theorem 5.1

The most important steps of the proof are outlined in the itemized list below.

(i) System (5.1)–(5.2) can be described in new coordinates (X, R, Ω) where $X := (\tilde{x}, \tilde{v}) := (x_{1j}, v_{1j})_{j \in \mathbf{2}:\mathbf{n}} \in \mathsf{X} := \mathbb{R}^{3(n-1)} \times \mathbb{R}^{3(n-1)}$ are translational states relative to robot 1, $R := (R_1, \dots, R_n) \in \mathsf{R} := \mathsf{SO}(3)^n$ is the collection of robot attitudes and $\Omega := (\Omega_1^1, \dots, \Omega_n^n) \in \Omega := \mathbb{R}^{3n}$ is the collection of robot angular velocities. The system dynamics in new coordinates are given by

$$\begin{aligned} \dot{x}_{1j} &= v_{1j}, \\ \dot{v}_{1j} &= -\frac{1}{m_j} R_j e_3 u_j + \frac{1}{m_1} R_1 e_3 u_1, \quad j \in \mathbf{2}:\mathbf{n}, \end{aligned} \tag{5.8}$$

$$\begin{aligned} \dot{R}_i &= R_i (\Omega_i^i)^\times, \\ J_i \dot{\Omega}_i^i &= \tau_i - \Omega_i^i \times J_i \Omega_i^i, \quad i \in \mathbf{n}, \end{aligned} \tag{5.9}$$

and the rendezvous manifold Γ expressed in new coordinates is

$$\tilde{\Gamma} = \{(X, R, \Omega) \in \mathsf{X} \times \mathsf{R} \times \Omega : X = 0,\ \Omega = \boldsymbol{\Omega}(X, R)\}. \tag{5.10}$$

(ii) The vector $y_i = (x_{ij}, v_{ij})_{j \in \mathcal{N}_i}$ is a linear function of X denoted $y_i = h_i(X)$ and $f_i(y_i)$ represented in new coordinates is denoted

$$g_i(X) := (f_i \circ h_i)(X). \tag{5.11}$$

Since $f_i(y_i)$ is a double integrator consensus controller for robot $i \in \mathbf{n}$, the origin $X = 0$ of system (5.5) in relative translational coordinates

$$\begin{aligned} \dot{x}_{1j} &= v_{1j}, \\ \dot{v}_{1j} &= g_j(X) - g_1(X), \quad j \in \mathbf{2}:\mathbf{n} \end{aligned} \tag{5.12}$$

is globally asymptotically stable. Since system (5.12) is linear, there exists a quadratic Lyapunov function $V : \mathsf{X} \to \mathbb{R}$, $V(X)$, whose Lie derivative along the vector field (5.12) is negative definite.

(iii) Use $V(X)$ to construct the following Lyapunov function for system (5.8)–(5.9) in (X, R, Ω) coordinates

$$W(X, R, \Omega) = \alpha W_{\text{tran}}(X) + W_{\text{rot}}(X, R, \Omega),$$

where $\alpha > 0$ is a parameter and

$$W_{\text{tran}}(X) = \sqrt{V(X)} + \frac{1}{2}V(X),$$

$$W_{\text{rot}}(X, R, \Omega) = \sum_{i=1}^{n} g_i^i(X, R) \cdot e_3 + \frac{1}{2}(\Omega - \boldsymbol{\Omega}(X, R))^\top \mathbb{J}(\Omega - \boldsymbol{\Omega}(X, R)).$$

The first term of $W_{\text{tran}}(X)$ is homogeneous of degree one, while the second is homogeneous of degree two with respect to X.

(iv) There exists $\alpha^\star > 0$ such that for all $\alpha > \alpha^\star$, $W(X, R, \Omega)$ is positive definite. This follows from the homogeneity properties of $W_{\text{tran}}(X)$ and employing similar arguments as in Proposition A.5 in the appendix. Moreover, the level sets of $W(X, R, \Omega)$ are compact and $W(0)^{-1} = \tilde{\Gamma}$.

(v) Drawing strongly upon the homogeneity properties of the Lie derivative of $W(X, R, \Omega)$ along the vector field (5.8)–(5.9) with respect to X, one can show that for any $\epsilon > 0$ there exist positive gains (k_1, k_2) and a compact level set W_δ of $W(X, R, \Omega)$ contained in an ϵ neighborhood of $\tilde{\Gamma}$ such that outside of the set W_δ, the lie derivative of $W(X, R, \Omega)$ is negative.

(vi) Applying Proposition A.1 in the appendix, it is shown that $\tilde{\Gamma}$ is globally practically stable.

□

5.4 Remarks on the Proposed Controller

Theorem 5.1 asserts global practical stability of the rendezvous manifold Γ. The reason that the stability is practical and not asymptotic is roughly as follows. In order to achieve rendezvous of the flying robots, $u_i q_i$ is driven approximately to $m_i f_i(y_i)$. What's important is not so much the difference in magnitude of these vectors but rather the difference in angle between them. In Fig. 5.2, one can see that $\boldsymbol{\Omega}_i$ acts to reduce this angle with a rate proportional to the magnitude of $\boldsymbol{\Omega}_i$. Since $\boldsymbol{\Omega}_i$ is a linear function of $f_i(y_i)$, as the robots approach consensus, $\boldsymbol{\Omega}_i$ converges to zero at the same rate as $f_i(y_i)$. This leads to increasing inaccuracy in closing the gap between the vectors $u_i q_i$ and $m_i f_i(y_i)$ insomuch that in a very small neighborhood of rendezvous, $\boldsymbol{\Omega}_i$ is so small that it fails to make the translational dynamics act as double integrators. More detailed reasoning is provided in the proof of Theorem 5.1 in [1].

We highlight below a list of interesting features of our control solution.

(i) Controller (5.7) is static, and it does not depend on auxiliary states that require communication between neighboring robots.
(ii) Controller (5.7) is local and distributed in the sense of Definition 2.2. Interestingly, it does not require sensing of relative attitudes, which can be computed using on-board cameras, but are harder to compute than relative displacements.
(iii) On the rendezvous manifold Γ, there is no prespecified thrust direction q_i for robot i and the robot thrust directions do not need to align at rendezvous. This is desirable if one wants to employ the proposed controller in a hierarchical control setting to enforce additional control specifications.
(iv) The proposed control law in this chapter does not guarantee hovering of the robots. While the robots converge to each other, nothing can be said about the motion of the ensemble. This cannot be otherwise, for it would be impossible to solve RP-F with hovering without additional sensors. One would need some measurement of the gravity vector, for example provided by a three-axis accelerometer. Our perspective is that the proposed solution of RP-F with strictly local and distributed feedbacks can serve as a layer in a hierarchy of higher-level control specifications such as hovering and path following. That said, if all agents can measure gravity, an alternative solution proposed in [2] allows the robots to hover in steady state or have a desired final thrust in the direction of a common inertial vector sensed by all robots. This solution, however, requires the computation of the derivatives of the reference thrust force $f_i(y_i)$, as well as the communication of the thrust control inputs $u_i^\star$ between neighboring robots. Note that these requirements are absent in the control solution presented in this chapter.

5.5 Simulation Results

We consider a group of five robots with the sensor digraph in Fig. 5.4. The robot masses and inertia matrices are: $m_1 = 3$ Kg, $m_2 = 3$ Kg, $m_3 = 3.4$ Kg, $m_4 = 3.2$ Kg, $m_5 = 3.2$ Kg and $J_1 := \text{diag}\,(0.13, 0.13, 0.04)$ Kg·m^2, as in [3], $J_2 = J_1,\ J_3 = 1.4J_1,\ J_4 = 1.2J_1,\ J_5 = 1.2J_1$. We pick $a_{ij} = 0.3$ for all $j \in \mathcal{N}_i$ and $\gamma = 30$. The control gains k_1 and k_2 in (5.7) are chosen to be $k_1 = 2$ and $k_2 = 0.45$. The initial conditions of the robots are shown in Table 5.1. The initial attitudes $R_i(0)$ of the robots are: up(right), side(ways) 1, side(ways) 2 and (upside)down, respectively, given by:

$$\begin{bmatrix} 1 & 0 & 0 \\ 0 & 1 & 0 \\ 0 & 0 & 1 \end{bmatrix}, \begin{bmatrix} 1 & 0 & 0 \\ 0 & 0 & -1 \\ 0 & 1 & 0 \end{bmatrix}, \begin{bmatrix} 1 & 0 & 0 \\ 0 & 0 & 1 \\ 0 & -1 & 0 \end{bmatrix}, \begin{bmatrix} 1 & 0 & 0 \\ 0 & -1 & 0 \\ 0 & 0 & -1 \end{bmatrix}.$$

Figure 5.5 shows the simulation without the presence of disturbances, while Fig. 5.6 shows the simulation when disturbances are present. The disturbances are: an addi-

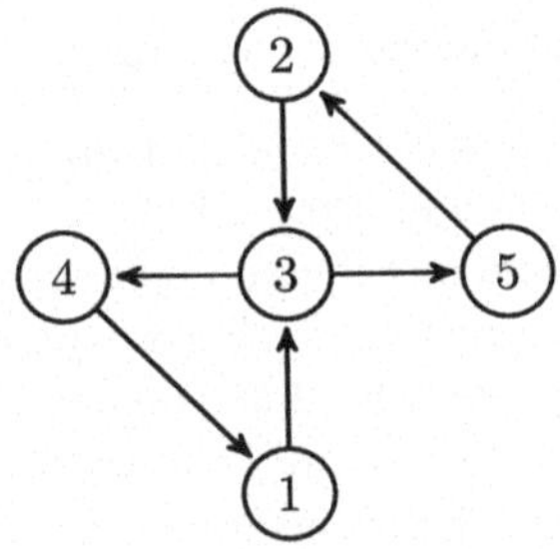

Fig. 5.4 Sensor digraph used in the simulation results. ©2017 IEEE. Reprinted, with permission, from [1]

Table 5.1 Simulation initial conditions ©2017 IEEE. Reprinted, with permission, from [1]

Vehicle i	$x_i(0)$ (m)	$v_i(0)$ (m/s)	$R_i(0)$
1	$(0, -10, 10)$	$(0, 0, 0)$	side 1
2	$(0, 10, 10)$	$(0, 0, 0)$	side 2
3	$(0, 0, 0)$	$(0, 0, 0)$	down
4	$(-10, 0, -10)$	$(0, 0, 0)$	up
5	$(10, 0, -10)$	$(0, 0, 0)$	up

tive random noise with maximum magnitude of 0.25 N on the applied force; an additive random noise with maximum magnitude of 0.25 N·m on the applied torque; an additive measurement error for the angular velocity, with maximum magnitude of 0.25 rad/s; an additive random noise on the quantity $f_i(y_i^i)$ accounting for errors in measurements of relative displacements and velocities of the vehicles. The direction of this vector has been rotated within 0.25 rad and the magnitude is scaled between 0.75 and 1.25 times the actual magnitude. The disturbances are updated 10 times per second. In both cases of Figs. 5.5 and 5.6, the vehicles' positions and velocities converge to a neighborhood of one another.

In Fig. 5.5, the vehicles remain within 0.25 m of one another, while in Fig. 5.6 the vehicles remain within 1m of one another at steady state. These neighborhoods can be made even smaller by further increasing the control gains k_1 and k_2. However, this would result in having higher control inputs. Metrics related to the thrust and torque inputs are presented in Table 5.2. The first two rows show peak control norms, and the last two show the root mean square (rms) of the control norms. In these simulations, we considered zero gravity, i.e., $g = 0$. This was done to improve visibility of the simulation results. In the presence of gravity, the vehicles would still converge to the same neighborhood of one another; however, at steady state, they would accelerate in the direction of gravity since gravity is not compensated through the control inputs in (5.7).

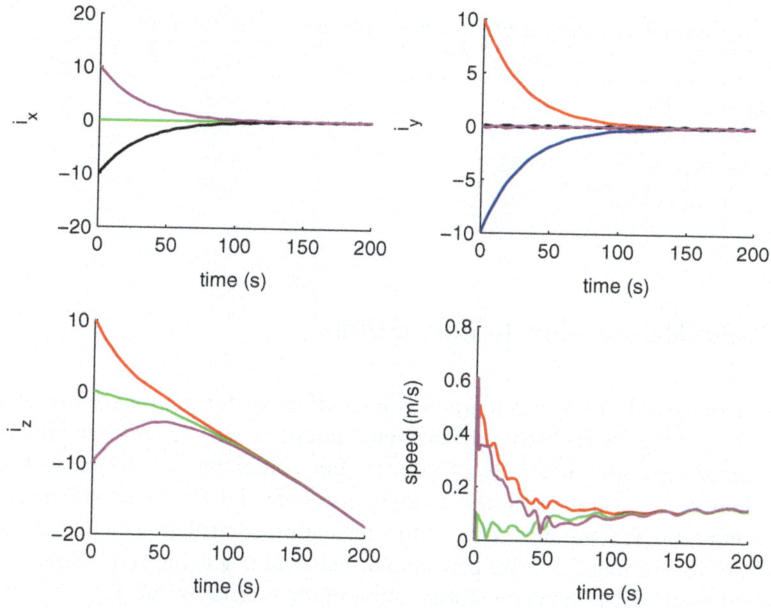

Fig. 5.5 Rendezvous control simulation without the presence of disturbances. At the top-left, top-right and bottom-left: positions of the five robots expressed in the inertial frame $\mathcal{I}$. At the bottom-right: linear speeds $\|v_i\|$, $i \in \mathbf{1:5}$. ©2017 IEEE. Reprinted, with permission, from [1]

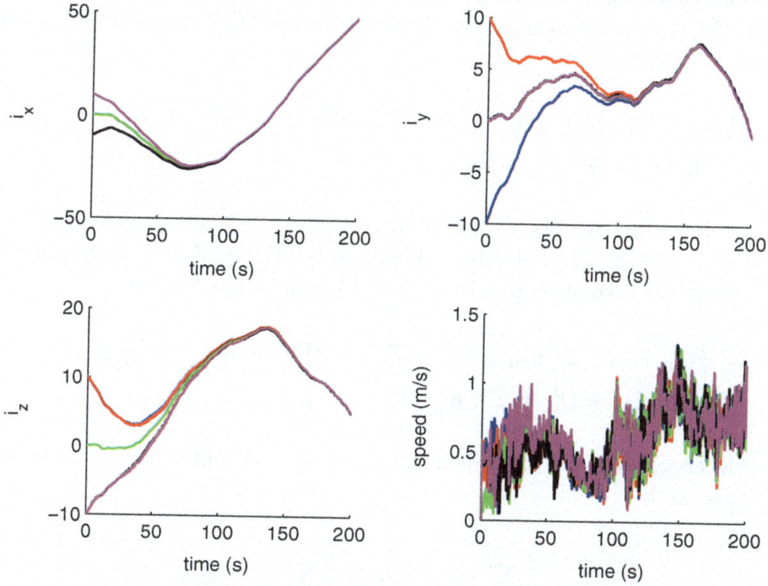

Fig. 5.6 Rendezvous control simulation with the presence of disturbances. At the top-left, top-right and bottom-left: positions of the five robots expressed in the inertial frame $\mathcal{I}$. At the bottom-right: linear speeds $\|v_i\|$, $i \in \mathbf{1:5}$. ©2017 IEEE. d, with permission, from [1]

Table 5.2 Control effort ©2017 IEEE. Reprinted, with permission, from [1]

	Figure 5.5	Figure 5.6
$\max_i \sup_t \lvert u_i(t)\rvert$ (N)	20.4	17.21
$\max_i \sup_t \lVert\tau_i(t)\rVert$ (N m)	15.27	16.47
$\max_i \mathrm{rms}(\lvert u_i(t)\rvert)$ (N)	1.72	4.31
$\max_i \mathrm{rms}(\lVert\tau_i(t)\rVert)$ (N m)	1.43	2.24

5.6 From Rendezvous to Formations

A notable omission in this monograph is a solution for formation control of flying robots. The analogous problem for kinematic unicycles will be presented in Chap. 7 using strictly local and distributed feedback. For completeness, in this section, we discuss how one can transform the rendezvous controller for flying robots in (5.7) (or the solution in [2] with hovering) into a formation controller. The final feedback, however, does not meet the strict local and distributed sensing requirements since each agent needs to know its own orientation in the inertial frame $\mathcal{I}$.

Consider any desired configuration of n flying robots $d = (d_1, \dots, d_n) \in \mathbb{R}^{3n}$ which, without loss of generality, is centered about the origin, i.e., $\sum_{i=1}^{n} d_i = 0$. A set of n flying robots is said to be in *formation*, when they satisfy the desired configuration d modulo translations; that is, the position of the ith robot satisfies $x_i = d_i + \bar{x}$ where $\bar{x} = (1/n)\sum_{i=1}^{n} x_i$ is the average position of the robots. Correspondingly, define the *formation manifold* as

$$\begin{aligned}\Gamma_f := \big\{&(x_i, v_i, R_i, \Omega_i^i)_{i\in\mathbf{n}} \in \mathbb{R}^{3n} \times \mathbb{R}^{3n} \times \mathsf{SO}(3)^n \times \mathbb{R}^{3n} : x_i = d_i + \bar{x},\\ &v_{ij} = 0,\ \Omega_i^i = \boldsymbol{\Omega}_i^i,\ i, j \in \mathbf{n}\big\}.\end{aligned} \tag{5.13}$$

For all $i \in \mathbf{n}$, define the fixed inertial vector $\delta_i := -d_i$ attached to robot i with endpoint $\hat{x}_i = x_i + \delta_i$. The collection of endpoints is denoted $\hat{x} := (\hat{x}_i)_{i\in\mathbf{n}}$ and $\hat{x}_{ij} := \hat{x}_j - \hat{x}_i$. Then, the formation manifold in (5.13) can be rewritten as

$$\begin{aligned}\Gamma_f := \big\{&(x_i, v_i, R_i, \Omega_i^i)_{i\in\mathbf{n}} \in \mathbb{R}^{3n} \times \mathbb{R}^{3n} \times \mathsf{SO}(3)^n \times \mathbb{R}^{3n} : \hat{x}_{ij} = 0,\\ &v_{ij} = 0,\ \Omega_i^i = \boldsymbol{\Omega}_i^i,\ i, j \in \mathbf{n}\big\}.\end{aligned} \tag{5.14}$$

This follows because $\hat{x}_{ij} = 0$ implies that $x_{ij} = d_j - d_i$ and therefore using the fact that $\sum_{i=1}^{n} d_i = 0$,

$$x_i - \bar{x} = \frac{1}{n}\sum_{j=1}^{n}(x_i - x_j) = \frac{1}{n}\sum_{j=1}^{n}(d_i - d_j) = d_i$$

as in (5.13). Therefore, Γ_f reduces to the rendezvous manifold in (3.1) in terms of $(\hat{x}_i, v_i, R_i, \Omega_i^i)_{i\in\mathbf{n}}$ quantities. It can be immediately shown that there is a diffeomorphism between $(x_i, v_i, R_i, \Omega_i^i)_{i\in\mathbf{n}}$ and $(\hat{x}_i, v_i, R_i, \Omega_i^i)_{i\in\mathbf{n}}$ where the latter can be treated as new states where $\dot{\hat{x}}_i$ satisfies

$$\dot{\hat{x}}_i = \dot{x}_i + \dot{\delta}_i = \dot{x}_i = v_i.$$

Therefore, the system equations in new coordinates are given by

$$\begin{aligned} \dot{\hat{x}}_i &= v_i, \\ m_i \dot{v}_i &= -u_i R_i e_3 + m_i g, \\ \dot{R}_i &= R_i \, (\Omega_i^i)^\times, \\ J_i \dot{\Omega}_i^i &= \tau_i - \Omega_i^i \times J_i \Omega_i^i. \end{aligned}$$

One achieves a formation controller that globally practically stabilizes Γ_f by replacing x_{ij}^i in (5.7) with $\hat{x}_{ij}^i$, satisfying

$$\hat{x}_{ij}^i = x_{ij}^i + (d_j - d_i)^i.$$

The term $(d_j - d_i)^i$ requires measurement of the fixed inertial vector $d_j - d_i$ in body frame. This, in turn, requires each agent to know its own attitude in the inertial frame. One could similarly adapt the rendezvous controller in [2] to a formation controller that allows the desired formation to hover, but requires communication of thrust inputs between neighboring robots.

References

1. Roza, A., Maggiore, M., Scardovi, L.: Local and distributed rendezvous of underactuated rigid bodies. IEEE Trans. Autom. Control **62**(8), 3835–3847 (2017)
2. Roza, A., Maggiore, M., Scardovi, L.: A class of rendezvous controllers for underactuated thrust-propelled rigid bodies. In: Proceedings of the 53rd IEEE Conference on Decision and Control, pp. 1649–1654 (2014)
3. Abdessameud, A., Tayebi, A.: Formation control of VTOL unmanned aerial vehicles with communication delays. Automatica **47**(11), 2383–2394 (2011)

Chapter 6
Rendezvous of Unicycles

Abstract In this chapter, we present a solution to the rendezvous control problem for kinematic unicycles (RP-U), in which the objective is to get a team of unicycles to convene at some common location by means of local and distributed feedback. The solution presented here is inspired by the rendezvous controller for flying robots developed in the previous chapter, but that result guaranteed global practical stability, whereas the feedbacks presented in this chapter globally asymptotically stabilize the rendezvous manifold.

6.1 Review of The Rendezvous Control Problem

Consider a team of n kinematic unicycles, whose model was derived in Chap. 2 and is given by

$$\begin{aligned} \dot{x}_i &= \begin{bmatrix} \cos(\theta_i) \\ \sin(\theta_i) \end{bmatrix} u_i, \\ \dot{\theta}_i &= \omega_i, \quad i \in \mathbf{n}. \end{aligned} \tag{6.1}$$

The schematic representation of one such robot is in Fig. 6.1. The pair $(x_i, \theta_i) \in \mathbb{R}^2 \times \mathbb{S}^1$ is the state of unicycle i, and the pair (u_i, ω_i), composed of the linear and angular speeds of unicycle i, is the control input.

The rendezvous problem for kinematic unicycles (RP-U) is similar to the rendezvous problem for flying robots presented in the previous chapter. Letting

$$\Gamma := \left\{ (x_i, \theta_i)_{i \in \mathbf{n}} \in \mathbb{R}^{2n} \times \mathbb{T}^n : x_{ij} = 0, \ i, j \in \mathbf{n} \right\}, \tag{6.2}$$

the goal of RP-U is to design local and distributed feedbacks $(u_i^\star, \omega_i^\star) : y_i^i \mapsto (u_i, \omega_i)$, $i \in \mathbf{n}$, rendering Γ globally asymptotically stable under the constraint $(u_i^\star, \tau_i^\star)|_\Gamma = (0, 0)$. We recall that the last requirement means that the unicycles must

A. Roza et al., *Distributed Coordination Theory for Robot Teams*, Lecture Notes in Control and Information Sciences 490,
https://doi.org/10.1007/978-3-030-96087-2_6

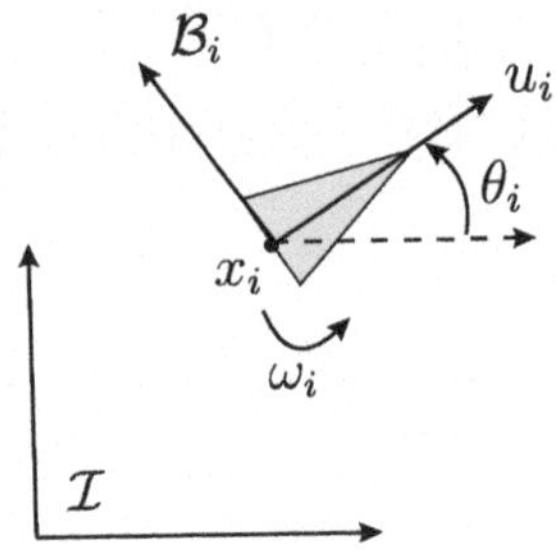

Fig. 6.1 Schematic representation of the kinematic unicycle model

stop when they achieve rendezvous. Notice further that, as in the previous chapter, the set Γ does not impose conditions on the relative heading angles of the agents.

6.2 Solution of the Rendezvous Control Problem

Our control solution makes use of the consensus controller for single integrators, a control primitive introduced in Chap. 4 given by

$$f_i(y_i) := \sum_{j \in \mathcal{N}_i} a_{ij} x_{ij}, \tag{6.3}$$

where $y_i = (x_{ij})_{j \in \mathcal{N}_i}$. We define $F_i(\mathcal{G}, \rho_1, \rho_2)$ as the set of integrator consensus controllers in (6.3) such that $a_{ij} > 0$ and $0 < \rho_1 < a_{ij} < \rho_2$ for all $j \in \mathcal{N}_i$ with sensor digraph $\mathcal{G}$.

The proposed control architecture resembles the one adopted in the previous chapter and is illustrated in the block diagram of Fig. 6.2. There are two nested loops. The outer loop treats each robot as a single integrator driven by the linear consensus controller (6.3)

$$\dot{x}_i = f_i(y_i), \; i \in \mathbf{n}. \tag{6.4}$$

Notice that the consensus set $\left\{(x_i)_{i \in \mathbf{n}} \in \mathbb{R}^{2n} : x_{ij} = 0, \; i, j \in \mathbf{n}\right\}$ is globally asymptotically stable for (6.4) if the sensing digraph has a globally reachable node [1] (see Chap. 4). The idea behind the design of the controller is similar to what we described in Chap. 5 when we addressed the rendezvous problem for flying robots, that is, to align the robot velocity to the consensus controller reference (6.3).

Analogous to the previous chapter, the signal $f_i(y_i(t))$ is computed in the body frame $\mathcal{B}_i$ and used as a reference signal for the inner-loop thrust and rotational controllers that assign the unicycle control inputs. The intuition behind these controllers is shown in Fig. 6.3 and explained below. The speed input u_i is the dot product

$$u_i = u_i^{\star}(y_i^i) = \|f_i(y_i^i)\| f_i(y_i^i) \cdot e_1,$$

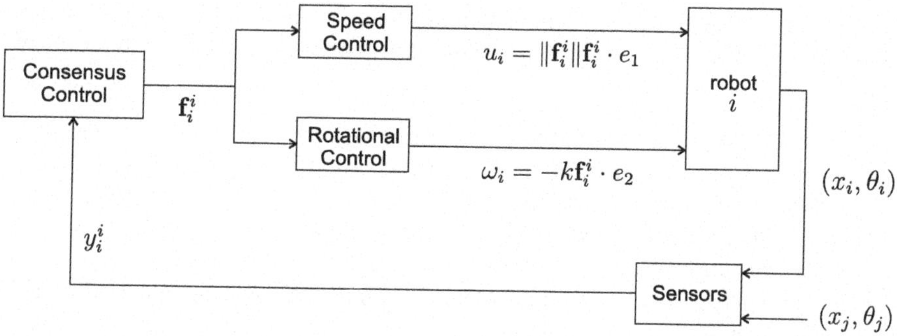

Fig. 6.2 Block diagram of the rendezvous control system for robot i. ©2018 IEEE. Reprinted, with permission, from [2]

Fig. 6.3 Illustration of the control inputs $u_i = u_i^\star(y_i^i)$ and $\omega_i = \omega_i^\star(y_i^i)$ in (6.5). ©2018 IEEE. Reprinted, with permission, from [2]

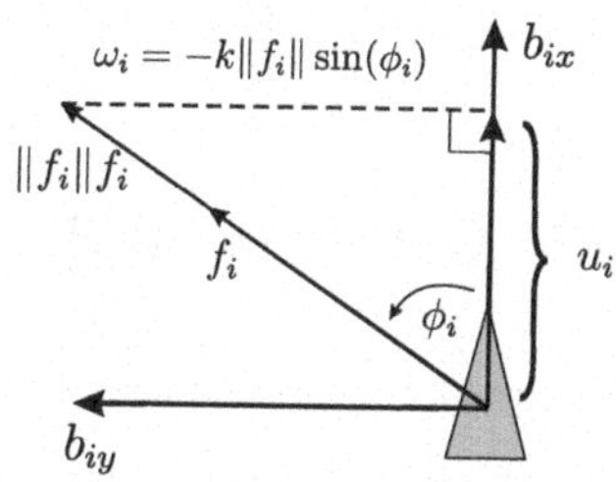

which is the projection of the reference $\|f_i(y_i)\|f_i(y_i)$ onto the heading axis $b_{ix} = [\cos(\theta_i), \sin(\theta_i]^T$ of robot i. The angular speed, on the other hand, is proportional to the dot product between the reference $f_i(y_i)$ and the second body axis $b_{iy} = [-\sin(\theta_i), \cos(\theta_i]^T$. In Fig. 6.3, one can see that

$$\omega_i = \omega_i^\star(y_i^i) = -kf_i(y_i^i) \cdot e_2 = -k\|f_i(y_i)\|\sin(\phi_i)$$

acts to reduce the angle ϕ_i between b_{ix} and $f_i(y_i)$ with a rate proportional to the magnitude of f_i. Together, these control inputs drive the robot velocity $u_i b_{ix}$ approximately to the reference $\|f_i(y_i)\|f_i(y_i)$.

The careful reader might argue that a more natural solution would have been to project $f_i(y_i)$ onto the heading axis b_{ix} of robot i instead of $\|f_i(y_i)\|f_i(y_i)$. The reason for the proposed choice is that, since $\|f_i(y_i)\|f_i(y_i)$ is homogeneous of degree two, $\omega_i^\star(y_i^i)$ converges to zero slower than $u_i^\star(y_i^i)$ as the robots approach consensus. This allows $\omega_i^\star(y_i^i)$ to exert sufficient control authority even as the robots converge to consensus, closing the gap between the vectors $u_i b_{ix}$ and $\|f_i(y_i)\|f_i(y_i)$. This property is key in achieving global asymptotic stability of the rendezvous manifold.

To summarize, the proposed feedback is

$$\begin{aligned} u_i &= u_i^\star(y_i^i) = \|f_i(y_i^i)\|f_i(y_i^i) \cdot e_1, \\ \omega_i &= \omega_i^\star(y_i^i) = -kf_i(y_i^i) \cdot e_2, \ i \in \mathbf{n}. \end{aligned} \tag{6.5}$$

The result below states that for sufficiently large k, the feedbacks in (6.5) solve RP-U if the network of unicycles has a sensor digraph containing a globally reachable node.

Theorem 6.1 RP-U *is solvable for system* (6.1) *if and only if the sensor digraph* $\mathcal{G}$ *contains a globally reachable node, in which case a solution is the following. For each positive integer* n*, and each* ρ_1, ρ_2 *such that* $0 < \rho_1 < \rho_2$*, there exists* $k^\star > 0$ *such that for all* $k > k^\star$*, for all sensor digraphs* $\mathcal{G}$ *with* n *nodes containing a globally reachable node, and for all linear functions* $f_i \in F_i(\mathcal{G}, \rho_1, \rho_2)$, $i \in \mathbf{n}$*, feedback* (6.5) *globally asymptotically stabilizes the rendezvous manifold* Γ *in* (6.2) *and solves* RP-U.

In the next section, we provide an outline of the proof of Theorem 6.1. For the full proof, we refer the reader to [2]. In Sect. 6.4, we make some remarks about the proposed controller, and in Sect. 6.5, we present simulation results.

6.3 Outline of the Proof of Theorem 6.1

The most important steps of the proof are outlined in the itemized list below.

(i) Start by considering the special case where the sensor graph $\mathcal{G}$ is strongly connected.

(ii) For system (6.1) consider new (X, θ) coordinates where $X := (X_i)_{i \in \mathbf{n}} \in \mathbb{R}^{2n}$,

$$X_i := f_i(y_i)/A_i, i \in \mathbf{n}, \tag{6.6}$$

are the new translational states lying in a subspace of $\mathbb{R}^{2n}$, with $A_i := \sum_{j \in \mathcal{N}_i} a_{ij}$ and $\theta := (\theta_i)_{i \in \mathbf{n}} \in \mathbb{T}^n$ contains all the robot attitudes.

(iii) The closed-loop unicycle dynamics of system (6.1) with feedback (6.5) in (X, θ) coordinates are given by

$$\dot{X}_i = \frac{\sum_{j \in \mathcal{N}_i} a_{ij}((\hat{g}_j^j \cdot e_1)R_j e_1 - (\hat{g}_i^i \cdot e_1)R_i e_1)}{A_i}, \tag{6.7}$$

$$\dot{\theta}_i = -k \hat{f}_i^i \cdot e_2, \tag{6.8}$$

where

$$\bar{a} := (a_{ij})_{(i,j) \in \mathcal{E}},$$
$$\hat{f}_i^i(\bar{a}, X_i, \theta_i) := A_i R_i^{-1}(\theta_i) X_i,$$
$$\hat{g}_i^i(\bar{a}, X_i, \theta_i) := A_i{}^2 R_i^{-1}(\theta_i) \|X_i\| X_i.$$

The rendezvous manifold Γ in new coordinates is given by

$$\tilde{\Gamma} := \{(X, \theta) \in \mathsf{X} \times \mathbb{T}^n : X = 0\}. \tag{6.9}$$

(iv) Consider the following functions

$$
\begin{aligned}
V(\gamma, X) &= \sum_{i \in \mathbf{n}} \gamma_i X_i^\top X_i, \\
W_{\text{tran}}(\gamma, X) &= \sqrt{V(\gamma, X)}, \\
W_{\text{rot}}(\bar{a}, X, \theta) &= \sum_{i \in \mathbf{n}} \hat{f}_i^i(\bar{a}, X_i, \theta_i) \cdot e_1,
\end{aligned}
\tag{6.10}
$$

where the gain $\gamma_i(\bar{a}) > 0$ is a continuous function of $\bar{a}$ which lies in a compact set. Use $W_{\text{tran}}(\gamma, X)$ and $W_{\text{rot}}(\bar{a}, X, \theta)$ to construct a Lyapunov function

$$
W(\gamma, \bar{a}, X, \theta) = \alpha W_{\text{tran}}(\gamma, X) + W_{\text{rot}}(\bar{a}, X, \theta) \tag{6.11}
$$

for the system in (X, θ) coordinates where $\alpha > 0$ is a design parameter. This function is homogeneous of degree 1 with respect to X.

(v) There exists $\alpha^\star > 0$ such that for all $\alpha > \alpha^\star$, $W(\gamma, \bar{a}, X, \theta)$ is positive semi-definite and $W(0)^{-1} = \tilde{\Gamma}$. This follows from Proposition A.5.

(vi) The Lie derivative of $W(X, \theta)$ along the vector field (6.7)–(6.8) is a homogeneous function of degree 2 with respect to X and is less than zero for all X in the set $\{X \in \mathbb{R}^{2n} : \|X\| = 1\}$.

(vii) Applying similar arguments as in Proposition A.4 in the appendix, the set $\tilde{\Gamma}$ is globally asymptotically stable. This step relies heavily on the homogeneity properties of $W(X, \theta)$ and its Lie derivative along the closed-loop vector field.

(viii) The proof can be extended from the class of strongly connected sensor graphs discussed so far to any graph containing a globally reachable node. If $\mathcal{G}$ is a sensor graph containing a globally reachable node, it can be decomposed into strongly connected components and has an associated condensation acyclic digraph $\mathcal{C}(\mathcal{G})$ as outlined in Sect. B.0.3.

(ix) The agents in the head node $\mathcal{V}_0 = \bar{\mathcal{L}}_0$ form an independently evolving system connected with a strongly connected graph. By the previous arguments, the agents in $\bar{\mathcal{L}}_0$ achieve rendezvous globally.

(x) Define the *reduced rendezvous manifold*

$$
\Gamma_{\bar{\mathcal{L}}_j} := \left\{ (x_i, \theta_i)_{i \in \bar{\mathcal{L}}_j} : x_{ik} = 0,\ i, k \in \bar{\mathcal{L}}_j \right\}.
$$

Suppose that, for some integer $j \geq 0$, the set $\Gamma_{\bar{\mathcal{L}}_{j-1}}$ is globally asymptotically stable for the system of unicycles in $\bar{\mathcal{L}}_{j-1}$. Then, there exists $l^\star > 0$ such that for any $k > l^\star$ in (6.5) and for all linear functions $f_i \in F_i(\mathcal{G}, \rho_1, \rho_2)$, $i \in \mathbf{n}$, feedback (6.5) globally asymptotically stabilizes $\Gamma_{\bar{\mathcal{L}}_j}$ for the system of unicycles in $\bar{\mathcal{L}}_j$. This is proved using the reduction theorem (Theorem A.5). For example, since $\Gamma_{\bar{\mathcal{L}}_0}$ is globally asymptotically stable, so is the set $\Gamma_{\bar{\mathcal{L}}_1}$.

(xi) Traversing the nodes in the condensation graph $\mathcal{C}(\mathcal{G})$ from head to tail, one can prove Theorem 6.1 using mathematical induction arguments.

□

6.4 Remarks on the Proposed Controller

Note that the lower bound on k is uniform over all sensor digraphs with n nodes, and all consensus controllers f_i with gains a_{ij} in a fixed compact interval $[\rho_1, \rho_2]$. In practice, this implies that one can tune the controller (6.5) without knowing the sensor digraph $\mathcal{G}$ nor the controller parameters a_{ij}. All that is required is to know bounds on these parameters. Moreover, it is clear that on the rendezvous manifold Γ, $y_i = 0$ for all $i \in \mathbf{n}$ and it follows that the requirement $(u_i^\star, \omega_i^\star)|_\Gamma = (0, 0)$ holds.

The feedback in (6.5) is very similar in form to the one in [3] given by,

$$\begin{aligned} u_i^\star(y_i^i) &= f_i(y_i^i) \cdot e_1, \\ \omega_i^\star(y_i^i) &= f_i(y_i^i) \cdot e_2, \ i \in \mathbf{n}. \end{aligned} \tag{6.12}$$

with $a_{ij} = 1$ for all $j \in \mathcal{N}_i$. The main difference in (6.5) is the extra multiplicative factor $\| f_i(y_i^i)\|$ in the control $u_i^\star(y_i^i)$ and the control gain k chosen sufficiently large. While the feedback in (6.12) achieves rendezvous for undirected and connected graphs, the solution cannot be generalized to the broad class of directed graphs considered in this chapter in that, as shown in [4], when the sensor digraph is a directed ring, the feedback (6.12) drives the unicycles to a circular formation instead of achieving rendezvous. Our solution in (6.5) can be viewed as an adaptation of the controller in (6.12) that allows for rendezvous with directed graph topologies containing a globally reachable node.

Moreover, in the presence of link failures in the sensor digraph, as long as the resulting graph after the last failure maintains a globally reachable node, the presented solution in (6.5) will still achieve rendezvous. It remains an open problem to solve RP-U with saturated feedback. If the rotational control gain $k > k^\star$ results in feedbacks that are too large, the magnitude of the gains a_{ij} can be reduced appropriately to avoid actuator saturation over any compact set of initial conditions. This, however, would slow the convergence rate.

6.5 Simulation Results

We consider a group of five robots with sensor digraph in Fig. 6.4. For the feedback in (6.5), we pick $a_{ij} = 0.05$ for all $j \in \mathcal{N}_i$. The control gain k is chosen to be $k = 0.2$.

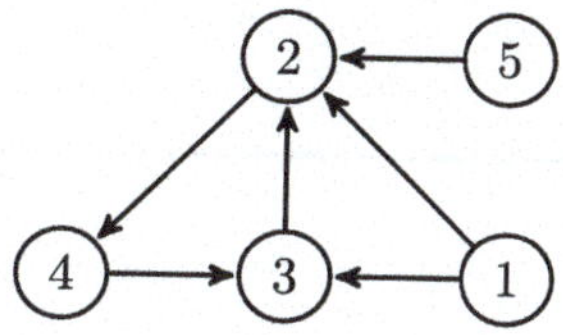

Fig. 6.4 Sensor digraph used in the simulation results. ©2018 IEEE. Reprinted, with permission, from [2]

Table 6.1 Simulation initial conditions ©2018 IEEE. Reprinted, with permission, from [2]

Vehicle i	$x_i(0)$ (m)	$\theta_i(0)$ (rad)
1	(0, 10)	0
2	$(-10, -10)$	$2\pi/5$
3	$(-50, 10)$	$4\pi/5$
4	$(-10, 0)$	$6\pi/5$
5	(10, 0)	$8\pi/5$

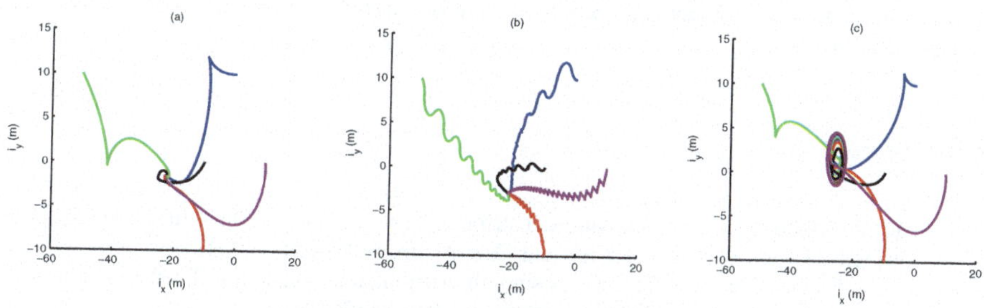

Fig. 6.5 Rendezvous control simulation for: **a** proposed feedback in (6.5), **b** feedback in [5], and **c** feedback in [3]. ©2018 IEEE. Reprinted, with permission, from [2]

The initial conditions of the robots are shown in Table 6.1. The simulation is presented in Fig. 6.5(a), and the control inputs for the five vehicles are plotted in Fig. 6.6 showing reasonable linear and rotational speeds. In Fig. 6.5 the trajectories corresponding to the feedback in [5], the feedback in [3] and our solution are compared. Notice that some curves in the figure have cusps although the actual solution is C^1. These cusps, however, are present only because the state is being projected into the inertial (i_x, i_y) plane; i.e., the angular states θ are not illustrated. The simulations are run with the initial conditions in Table 6.1 and the sensing digraph in Fig. 6.4. It can be seen that the proposed feedback has practical advantages over the time-varying feedback in [5]. The proposed feedback induces a more natural behavior in the ensemble of unicycles. The feedback in [5] makes the unicycle "wiggle" indefinitely, a behaviour which would be unacceptable in practice. The feedback in [3] causes the unicycles to converge to a circular formation instead of rendezvous. Rendezvous is guaranteed only for undirected graphs in this case.

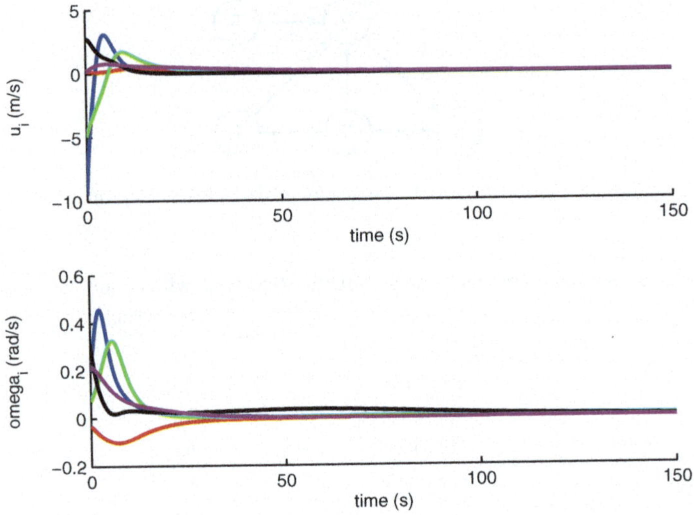

Fig. 6.6 Simulation control inputs for proposed feedback in (6.5). ©2018 IEEE. Reprinted, with permission, from [2]

References

1. Moreau, L.: Stability of continuous-time distributed consensus algorithms. In: Proceedings of the 43rd IEEE Conference on Decision and Control, pp. 3998–4003 (2004)
2. Roza, A., Maggiore, M., Scardovi, L.: A smooth distributed feedback for global rendezvous of unicycles. IEEE Trans. Control Netw. Syst. **5**(1), 640–652 (2018)
3. Zheng, R., Lin, Z., Cao, M.: Rendezvous of unicycles with continuous and time-invariant local feedback. In: Proceedings of the 18th IFAC World Congress, pp. 10044–10049. Milano, Italy (2011)
4. Zheng, R., Lin, Z., Yan, G.: Ring-coupled unicycles: boundedness, convergence, and control. Automatica **45**(11), 2699–2706 (2009)
5. Lin, Z., Francis, B., Maggiore, M.: Necessary and sufficient graphical conditions for formation control of unicycles. IEEE Trans. Autom. Control **50**(1), 121–127 (2005)

Chapter 7
Unicycle Formations Coming to Rest

Abstract In this chapter, we present a control solution to the Parallel Formation Problem (PP), in which the objective is to make a team of unicycles achieve a formation with parallel headings and come to a stop. As a special case of PP, we discuss parallel line formations and full synchronization of unicycles.

7.1 The Parallel Formation Problem (PP)

Consider n unicycles

$$\begin{aligned}\dot{x}_i &= \begin{bmatrix}\cos(\theta_i)\\ \sin(\theta_i)\end{bmatrix} u_i,\\ \dot{\theta}_i &= \omega_i, \quad i \in \mathbf{n}.\end{aligned} \tag{7.1}$$

As introduced in Chap. 3, a formation is a geometric pattern defined modulo roto-translations. In this section, we consider the subset of parallel formations in which the headings of all unicycle agents in the team align with one another.

For the reader's convenience, we recall the main setup presented in Chap. 3. Let $d \in \mathbb{R}^{2(n-1)}$ be a formation vector. Without loss of generality, we choose unicycle 1 to be at the front of the formation so that $d^1_{1i} \cdot e_1 \leq 0$ for all $i \in \mathbf{2}\!:\!\mathbf{n}$. The set of all formation vectors is

$$\mathsf{F} := \{d \in \mathbb{R}^{2(n-1)} : d^1_{1i} \cdot e_1 \leq 0, \ i \in \mathbf{2}\!:\!\mathbf{n}\},$$

and the formation manifold is given by

$$\Gamma := \left\{(x, \theta) \in \mathbb{R}^{2n} \times \mathbb{T}^n : x_{1i} = R_1 d^1_{1i}, \ i \in \mathbf{n}\right\}.$$

A. Roza et al., *Distributed Coordination Theory for Robot Teams*, Lecture Notes in Control and Information Sciences 490,
https://doi.org/10.1007/978-3-030-96087-2_7

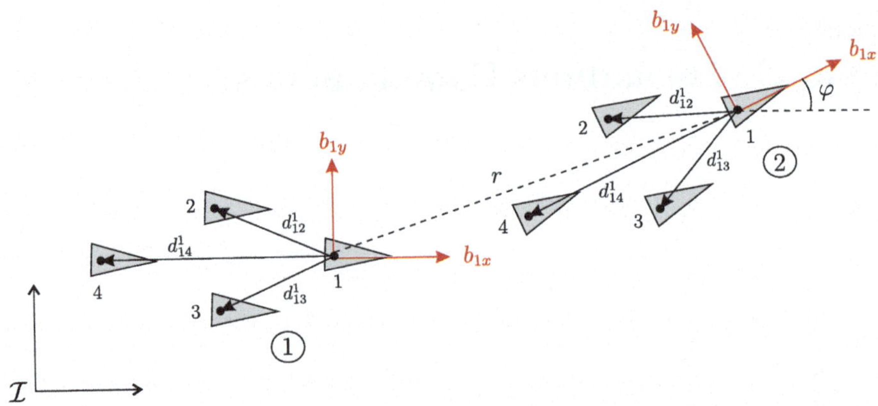

Fig. 7.1 Formation in terms of fixed relative displacement vectors d_{1i}^1, $i \in \mathbf{2}:\mathbf{n}$

The *parallel formation manifold*

$$\Gamma_p := \{(x, \theta) \in \Gamma : \theta_i = \theta_1, \ i \in \mathbf{2}:\mathbf{n}\} \tag{7.2}$$

is the subset of the state space in which the unicycles have parallel headings, and their relative displacements meet the formation specification. A parallel formation is illustrated in Fig. 7.1. For any $d \in \mathsf{F}$, the objective of the Parallel Formation Control Problem is to design local and distributed feedbacks rendering the parallel formation manifold almost semi-global asymptotically stable (ASGAS) under the constraint that $(u_i^\star, \omega_i^\star)|_{\Gamma_p} = (0, 0)$ for all agents $i \in \mathbf{n}$. This constraint means that the formation, once achieved, will stop. A formal problem statement has been presented in Sect. 3.3.1.

7.2 Solution of the Parallel Formation Problem

Our first step is to reformulate the Parallel Formation Problem as a consensus problem on a new set of variables.

Let $\bar{\alpha} > 0$ be a design parameter, and $d = (d_{i1}^1)_{i \in \{2,\ldots,n\}} \in \mathsf{F}$ be the desired parallel formation. Based on the formation also define the following values

$$\begin{aligned} &\alpha_1 := \bar{\alpha}, \ \beta_1 := 0, \\ &\alpha_i := -d_{1i}^1 \cdot e_1 + \bar{\alpha}, \ \beta_i := -d_{1i}^1 \cdot e_2, \ \ i \in \mathbf{2}:\mathbf{n}. \end{aligned} \tag{7.3}$$

Referring to Fig. 7.2, we attach the offset vector $\delta_i := \alpha_i R_i e_1 + \beta_i R_i e_2$ to the origin of unicycle i, and let $\hat{x}_i := x_i + \delta_i$ be the endpoint position of the offset vector in the coordinates of frame $\mathcal{I}$. Note that the offset vector δ_i has been defined

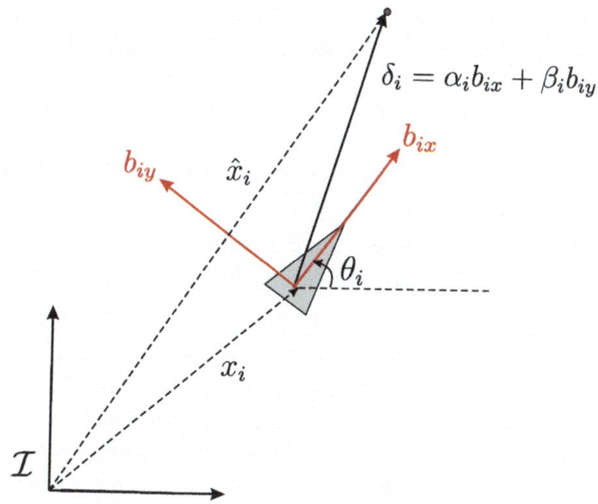

Fig. 7.2 Representation of the offset vector δ_i. ©2019 IEEE. Reprinted, with permission, from [1]

in terms of the two body axes $R_i e_1$ and $R_i e_2$ of unicycle i and has a body frame representation given by $\delta_i^i := \alpha_i e_1 + \beta_i e_2$. For each agent $i \in \mathbf{n}$, we also define the relative displacements of its endpoint $\hat{x}_i$ with respect to the endpoints of neighboring unicycles

$$\hat{x}_{ij} := \hat{x}_j - \hat{x}_i,$$

and collect all relative endpoint displacements pertaining to unicycle i in a vector $\hat{y}_i$,

$$\hat{y}_i := (\hat{x}_{ij})_{j \in \mathcal{N}_i}.$$

We now show that PP reduces to synchronizing the unicycles' heading angles θ_i and the endpoints $\hat{x}_i$ for all $i \in \mathbf{n}$. To this end, suppose that $\theta_{ij} = 0$ and $\hat{x}_{ij} = 0$ for all $i, j \in \mathbf{n}$. Then,

$$\begin{aligned} 0 = \hat{x}_{1i}^i &= [(x_i + \delta_i) - (x_1 + \delta_1)]^i \\ &= x_{1i}^i + (\delta_i - \delta_1)^i \\ &= x_{1i}^i - d_{1i}^1 = x_{1i}^1 - d_{1i}^1. \end{aligned}$$

The last identity follows from the fact that $R_i = R_1$. We conclude that $\theta_{ij} = 0$ and $\hat{x}_{ij} = 0$ for all $i, j \in \mathbf{n}$ implies $x_{1i}^1 = d_{1i}^1$, so that the unicycles satisfy the parallel formation requirement. Vice versa, it is clear that if the unicycles form a parallel formation, then $\theta_{ij} = 0$ and $\hat{x}_{ij} = 0$ for all $i, j \in \mathbf{n}$.

Having shown that PP amounts to the simultaneous synchronization of the headings θ_i and the endpoints $\hat{x}_i$, we now present feedbacks that do just that. Let $f_i(\cdot)$ be a uniformly bounded integrator consensus controller defined in (4.3) and $g_i(\cdot, \cdot)$ be an attitude synchronizer defined in (4.15). The basic idea behind the design of our control solution is to combine $f_i(\cdot)$ and $g_i(\cdot, \cdot)$ to achieve simultaneous synchronization of the headings θ_i and the endpoints $\hat{x}_i$. It turns out that the next controller achieves both these, at times competing, objectives simultaneously

$$\begin{aligned} u_i =& u_i^\star(y_i^i, \varphi_i) = f_i(\hat{y}_i^i) \cdot e_1 + \beta_i \omega_i^\star(y_i^i, \varphi_i), \\ \omega_i =& \omega_i^\star(y_i^i, \varphi_i) = \frac{1}{\alpha_i} \left(f_i(\hat{y}_i^i) \cdot e_2 + k g_i(\varphi_i, \eta) \right), \ i \in \mathbf{n}, \end{aligned} \tag{7.4}$$

where $k > 0$ is a high-gain parameter, and $\eta = (\eta_1, \ldots, \eta_n)$, $\eta_i := 1/\alpha_i$.

We are now ready to present the main result of this chapter.

Theorem 7.1 *Consider the collection of n unicycles in* (7.1) *with controller* (7.4), *where the functions* $f_i(\cdot)$, $g_i(\cdot)$ *are defined in* (4.3), (4.15) *and satisfy properties A1, A2 and B1–B3. Assume that sensor graph* $\mathcal{G}$ *is undirected and connected. For any parameters* $a_{ij} = a_{ji} > 0$ *in* (4.3) *and any parameters* $b_{ij} = b_{ji} > 0$ *in* (4.15) *satisfying* $(b_{ij})_{(i,j)\in\mathcal{E}} \in (\mathbb{R}_+)^{|\mathcal{E}|} \backslash N_b$ *as in Theorem* 4.2, *there exists* $\alpha^\star > 0$ *such that for any parallel formation* $d = (d_{1i}^1)_{i \in \mathbf{2:n}} \in \mathsf{F}$, *choosing* $\bar{\alpha} > \alpha^\star \max_{i\in\mathbf{2:n}} (-d_{1i}^1 \cdot e_1)$ *in* (7.3), *the formation manifold* Γ_p *is* **ASGAS** *with high-gain parameter k.*

Below we provide the reader with an outline of the proof of Theorem 7.1. For the full proof, the reader is referred to our paper [1]. Roughly speaking, the theorem states that letting the offset $\bar{\alpha}$ in (7.3) grow proportionally to the length of the formation (the quantity $\max_i (-d_{1i}^1 \cdot e_1)$), and choosing k in (7.4) to be sufficiently large, the controller (7.4) ensures that almost all initial conditions in any given compact set are contained in the domain of attraction of the formation manifold Γ_p. Another property of controller (7.4) is that $(u_i^\star, \omega_i^\star)\big|_{\Gamma_p} = 0$ for all $i \in \mathbf{n}$, and therefore the unicycles come to a stop as Γ_p is approached, as required in the statement of PP. In the next section, we further discuss the controller (7.4).

7.3 Outline of the Proof of Theorem 7.1

The most important steps of the proof are outlined in the itemized list below.

(i) Express system (7.1) in terms of new coordinates $(\hat{x}_i, \theta_i)_{i\in\mathbf{n}}$

$$\begin{aligned} \dot{\hat{x}}_i &= f_i(\hat{y}_i) + k g_i(\varphi_i, \eta) R_i e_2, \\ \dot{\theta}_i &= \frac{1}{\alpha_i} \left(f_i(\hat{y}_i) \cdot R_i e_2 + k g_i(\varphi_i, \eta) \right), \quad i \in \mathbf{n}. \end{aligned} \tag{7.5}$$

The parallel formation manifold Γ_p in (7.2) written in new coordinates becomes

$$\hat{\Gamma}_p := \left\{ (\hat{x}, \theta) \in \mathbb{R}^{2n} \times \mathbb{T}^n : \hat{x}_{1i} = 0,\ \theta_{1i} = 0,\ i \in \mathbf{n} \right\}. \tag{7.6}$$

(ii) Design translational and rotational storage functions, $V_t(x)$ and $V_r(\theta)$, respectively, as follows:

- Consider system (4.1) of integrators with bounded consensus feedback control (4.3), where $f(s)$ satisfies assumptions A1 and A2. This is a gradient system. Let $V_t(x)$ be the corresponding nonnegative storage function given by

$$V_t(x) = \frac{1}{2}\sum_{i=1}^{n}\sum_{j\in\mathcal{N}_i} a_{ij} \int_0^{\|x_{ij}\|} f(s)\mathrm{d}s \tag{7.7}$$

satisfying $\dot{x} = -\nabla V_t(x)$.

- Consider system (4.13) of rotational integrators with consensus feedback control (4.15) satisfying assumptions B1–B3. Inspired by [2] in the proof of Theorem 4.2, let $V_r(\theta)$ be the following storage function

$$V_r(\theta) := \frac{1}{2}\sum_{i=1}^{n}\sum_{j\in\mathcal{N}_i} b_{ij} \int_0^{\theta_{ij}} g(s)\mathrm{d}s. \tag{7.8}$$

(iii) Design a Lyapunov function V for the closed-loop system (7.5), by combining $V_t(\hat{x})$ and $V_r(\theta)$ as follows:

$$V(\hat{x}, \theta) := V_t(\hat{x}) + kV_r(\theta). \tag{7.9}$$

Notice that V is nonnegative and $V^{-1}(0) = \hat{\Gamma}_p$.

(iv) The Lie derivative of $V(\hat{x}, \theta)$ along the vector field (7.5) is less than or equal to zero. For sufficiently large $\bar{\alpha} > 0$, the zero level set of the lie derivative coincides with the formation manifold $\hat{\Gamma}_p$ on a neighborhood of $\hat{\Gamma}_p$. Via Lyapunov's direct method (see Theorem A.1 in the appendix), $\hat{\Gamma}_p$ is asymptotically stable.

(v) A further Lyapunov analysis is employed to show that $\hat{\Gamma}_p$ is in fact almost semiglobally asymptotically stable with high-gain parameter k.

□

7.4 Remarks on the Proposed Controller

As we mentioned earlier, the philosophy behind controller (7.4) is to convert PP into a synchronization problem in which we make the offset vectors $\hat{x}_i$ and the heading angles θ_i converge to one another.

The block diagram in Fig. 7.3 summarizes the design of feedbacks $(u_i, \omega_i)_{i\in\mathbf{n}}$. From its sensors, unicycle i obtains the vector (y_i^i, φ_i) of its heading and displacement relative to its neighbors. These quantities can be measured locally in unicycle i's body

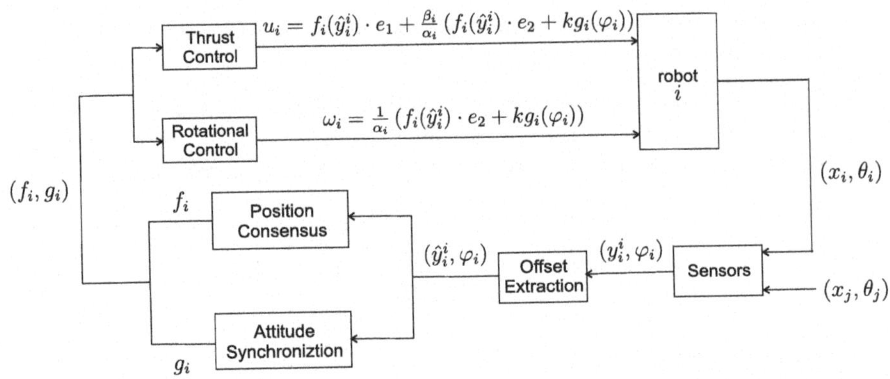

Fig. 7.3 Block diagram of the formation control system for robot i. ©2019 IEEE. Reprinted, with permission, from [1]

frame using, for example, on-board cameras. The offset extraction block takes as an input the vector (y_i^i, φ_i) and outputs $(\hat{y}_i^i, \varphi_i)$, where each component of $\hat{y}_i^i = (\hat{x}_{ij}^i)_{j\in\mathcal{N}_i}$ is computed as,

$$\hat{x}_{ij}^i = x_{ij}^i + \alpha_j R_j^i e_1 + \beta_j R_j^i e_2 - \begin{bmatrix}\alpha_i & \beta_i\end{bmatrix}^\top. \tag{7.10}$$

This computation requires that, in addition to (y_i^i, φ_i), unicycle i has access to the formation parameters $(\alpha_j, \beta_j)_{j\in\mathcal{N}_i}$ (defined in (7.3)) of its neighbors. These quantities must be stored in memory on-board unicycle i before deployment. Moreover, in order to compute $\hat{x}_{ij}^i$ in (7.10), unicycle i must be able to identify its neighbors so as to use, for each $j \in \mathcal{N}_i$, the appropriate bias constants (α_j, β_j). Such identification can be achieved, for instance, by means of visual markers. A consequence of using the constants (α_j, β_j) is that the unicycle feedbacks are not identical and the formation is not invariant to a relabelling of the agents. This is hardly surprising because, in our formulation of PP, we allow for general, nonsymmetric formations.

An important property of the feedback in (7.4) is that it is local and distributed, since $u_i^\star$ and $\omega_i^\star$ depend on (y_i^i, φ_i). As a consequence of this feature, the asymptotic position and orientation of the formation with respect to the inertial frame depend only on the initial configuration of the unicycles.

7.5 Special Cases: Line Formations and Full Synchronization

As a by-product of the formation control solution presented in the previous section, we derive corresponding solutions for the special cases of parallel line formations and full synchronization.

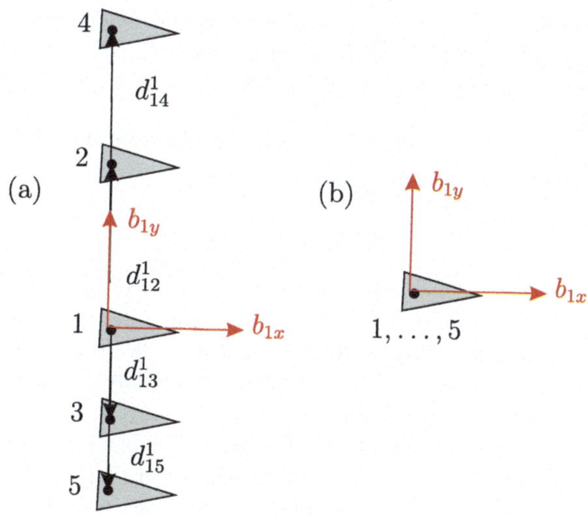

Fig. 7.4 **a** shows an example of a parallel line formation while **b** shows an example of full synchronization, a special case of a parallel line formation

A *parallel line formation* is a parallel formation where the robot's positions are aligned. Formally, we define

$$\mathsf{LF} := \{d \in \mathbb{R}^{2(n-1)} : d^1_{1i} \cdot e_1 = 0,\ i \in \mathbf{2}:\mathbf{n}\}.$$

Clearly, $\mathsf{LF} \subset \mathsf{F}$.

In the case of *full synchronization*, the unicycles have the same position and orientation with respect to the inertial frame, i.e., $d^1_{1i} = 0$ for all $i \in \mathbf{2}:\mathbf{n}$ (and therefore $\alpha_i = \bar{\alpha}$ and $\beta_i = 0$ for all $i \in \mathbf{n}$). Full synchronization, therefore, corresponds to the formation $0 \in \mathsf{F}$. Examples of a parallel line formation and full synchronization are illustrated in Fig. 7.4.

According to Theorem 7.1, in both of these cases, it suffices that $\bar{\alpha}$ satisfies the less strict condition $\bar{\alpha} > 0$. This is advantageous, as it will be discussed in Chap. 9 that large values of $\bar{\alpha}$ can slow down the rate of convergence of the unicycles to the formation. Arbitrarily choosing $\bar{\alpha} = 1$, the corresponding controller in (7.4) reduces to

$$\begin{aligned} u_i^\star(y_i^i, \varphi_i) &= f_i(\hat{y}_i^i) \cdot e_1 + \beta_i \omega_i^\star(y_i^i, \varphi_i), \\ \omega_i^\star(y_i^i, \varphi_i) &= f_i(\hat{y}_i^i) \cdot e_2 + k g_i(\varphi_i, \mathbf{1}),\ i \in \mathbf{n}, \end{aligned} \tag{7.11}$$

in which, $\hat{x}^i_{ij} = x^i_{ij} + R^i_j e_1 - e_1 + \beta_j R^i_j e_2 - \beta_i e_2$. Since the values $\alpha_i = \bar{\alpha} = 1$ for all $i \in \mathbf{n}$ are equal, unicycle i only needs to store the quantities $(\beta_j)_{j \in \mathcal{N}_i}$ of its neighbors on-board. The next corollary is a specialization of Theorem 7.1 to parallel line formations.

Corollary 7.1 *Consider the collection of n unicycles in* (7.1) *with controller* (7.11), *where the functions $f_i(\cdot)$, $g_i(\cdot)$ are defined as in (4.3), (4.15) and enjoy properties A1, A2 and B1–B3. Assume that sensor graph $\mathcal{G}$ is undirected and connected. For any* *parameters $a_{ij} = a_{ji} > 0$ in (4.3), any parameters $b_{ij} = b_{ji} > 0$ in (4.15) satisfying $(b_{ij})_{(i,j)\in\mathcal{E}} \in (\mathbb{R}_+)^{|\mathcal{E}|}\backslash N_b$ as in Theorem 4.2, and any parallel line formation $d \in$ LF, the formation manifold Γ_p is **ASGAS** with high-gain parameter k.*

In the special case of full synchronization, $\beta_i = 0$ for all $i \in \mathbf{n}$, and the controller in (7.11) reduces to

$$\begin{aligned} u_i^\star(y_i^i, \varphi_i) &= f_i(\hat{y}_i^i) \cdot e_1, \\ \omega_i^\star(y_i^i, \varphi_i) &= f_i(\hat{y}_i^i) \cdot e_2 + kg_i(\varphi_i, \mathbf{1}),\ i \in \mathbf{n}, \end{aligned} \tag{7.12}$$

in which $\hat{x}_{ij}^i = x_{ij}^i + R_j^i e_1 - e_1$. Since the α_i and β_i parameters are equal for all agents, unicycle i does not need to store any parameters of its neighbors on-board, the control inputs are identical for all unicycles, so that in this case the configuration is invariant to relabelling of agents. This is hardly surprising since the formation is symmetric in this case. The controller in (7.12) can be viewed as an extension of the result for unicycle rendezvous in Chap. 6. In fact, in Chap. 6 the controller was defined as

$$\begin{aligned} u_i^\star(y_i^i, \varphi_i) &= \|f_i(y_i^i)\| f_i(y_i^i) \cdot e_1, \\ \omega_i^\star(y_i^i, \varphi_i) &= -kf_i(y_i^i) \cdot e_2,\ i \in \mathbf{n}, \end{aligned} \tag{7.13}$$

in which $f_i(y_i^i) = \sum_{j\in\mathcal{N}_i} a_{ij} x_{ij}^i$ is a linear single integrator consensus controller.

While the controller in (7.13) guarantees global rendezvous, in which only the unicycle positions are synchronized, the controller in (7.12) guarantees almost semiglobal full synchronization where both positions and angles of the unicycles are synchronized. The control inputs in (7.12) and (7.13) are similar in structure. The main difference is that the full synchronization controller in (7.12) has an additional term $kg_i(\varphi_i, \eta)$ responsible for aligning the unicycle heading angles, not required for rendezvous. In fact, for unicycle i, (7.12) depends on (y_i^i, φ_i) while (7.13) depends only on y_i^i.

7.6 Simulation Results

This section presents simulations for a group of five unicycles to illustrate our results. The interaction function $f(s)$ for the bounded integrator consensus control is chosen as in (4.4) while the interaction function for the attitude synchronizer is chosen satisfying assumptions B1, B2 and B3 as in Fig. 4.1. The undirected sensing graph is cyclic with connections as shown in Fig. 7.5, and the desired triangular formation is specified by $d_{12}^1 = (-10, 5)$, $d_{13}^1 = (-10, -5)$, $d_{12}^1 = (-20, 10)$ and $d_{12}^1 = (-20, -10)$, as illustrated in Fig. 7.6. We have chosen random initial unicycle positions on a 40 m $\times$ 40 m area with random initial angles. The corresponding plot of a simulation run is shown in Fig. 7.7.

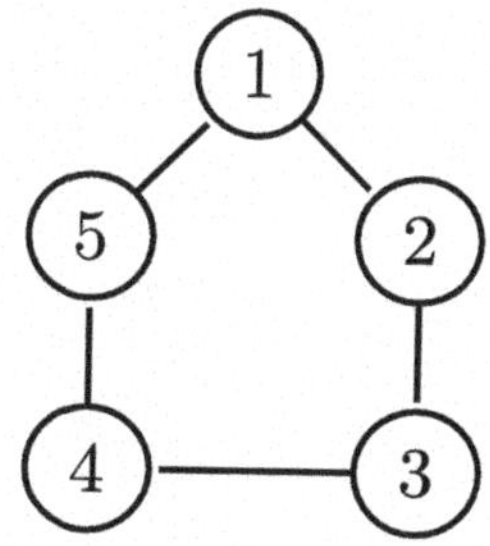

Fig. 7.5 Undirected graph $\mathcal{G}$ under consideration in the simulation results. ©2019 IEEE. Reprinted, with permission, from [1]

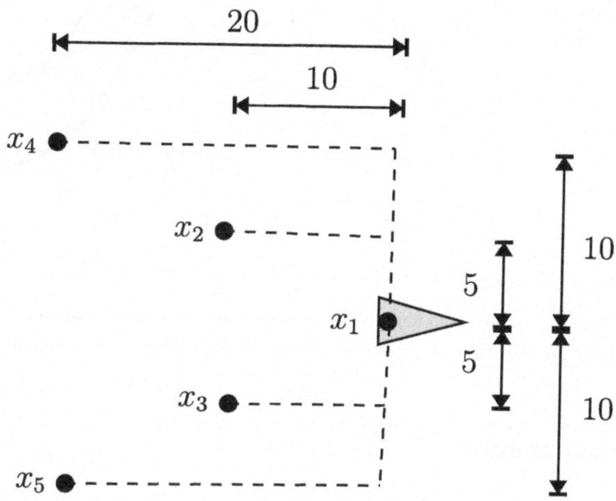

Fig. 7.6 Triangular formation specified by the offset vectors $d^1_{12} = (-10, 5)$, $d^1_{13} = (-10, -5)$, $d^1_{14} = (-20, 10)$ and $d^1_{15} = (-20, -10)$. ©2019 IEEE. Reprinted, with permission, from [1]

Extensive simulation trials will be presented in Chap. 9 to study the effectiveness of our control solution under different realistic scenarios not captured by the main result in Theorem 7.1 including:

- performance in the presence of state dependent sensor graphs in which each unicycle's neighbors are those that lie within a given radius of itself;
- performance for directed sensing graphs as opposed to undirected sensing graphs;
- performance when the high gain conditions on $\bar{\alpha}$ and k are ignored;
- robustness of the approach to unmodelled effects including sensor noise, input noise, sampling and saturated inputs.

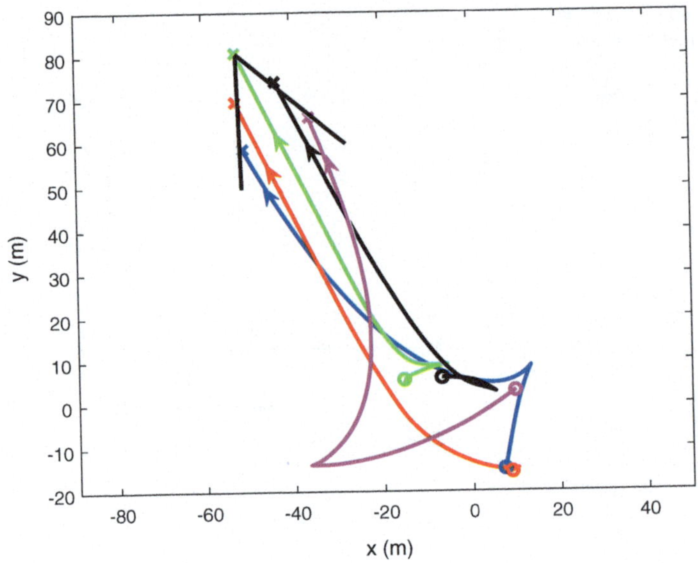

Fig. 7.7 Simulation for a triangle formation. Initial positions are indicated with ∘ and final positions are indicated with ×. ©2019 IEEE. Reprinted, with permission, from [1]

References

1. Roza, A., Maggiore, M., Scardovi, L.: A smooth distributed feedback for formation control of unicycles. IEEE Trans. Autom. Control **64**(12), 4998–5011 (2019)
2. Mallada, E., Freeman, R., Tang, A.: Distributed synchronization of heterogeneous oscillators on networks with arbitrary topology. IEEE Trans. Control Netw. Syst. **3**(1), 12–23 (2016)

Chapter 8
Unicycle Formations with Parallel and Circular Motions

Abstract Having presented, in Chap. 7, a local and distributed feedback making a team of unicycles achieve a formation and come to a stop, in this chapter, we take the next step and consider two kinds of final motions, parallel and circular. At the same time, we address two collective goals, formation path following and formation flocking.

8.1 Introduction

In this chapter, we continue our investigation of unicycle formations started in Chap. 7. However, while in Chap. 7 we investigated the simplest case of formations that come to a stop, in this chapter, we consider parallel and circular final motions, and for the collective goal, we investigate both formation path following and formation flocking. In this section, we briefly review the formulation of the various problems under consideration. First off, the following table provides a bird's-eye-view of the coordination problems discussed in this chapter. For convenience, we also recall the list of relevant abbreviations and target sets that need to be stabilized.

Final motion	Collective goal		
	Full sync	Formation path following	Formation flocking
Parallel	PPP,PFP,PFP-B	PPP	PFP,PFP-B
Circular	CPP,CFP	CPP	CFP

Final motion	Problem	Set
Linear	PFP: **P**arallel formation **F**locking **P**roblem	Γ_{pf}
	PFP-B: **P**arallel formation **F**locking **P**roblem w/ **B**eacon	Γ_{pfb}
	PPP: **P**arallel formation **P**ath following **P**roblem	Γ_{lp}
Circular	CFP: **C**ircular formation **F**locking **P**roblem	Γ_{cf}
	CPP: **C**ircular formation **P**ath following **P**roblem	Γ_{cp}

The "Full sync" column in the foregoing table refers to the special case when the formation vector d is chosen to be zero.

A. Roza et al., *Distributed Coordination Theory for Robot Teams*, Lecture Notes in Control and Information Sciences 490,
https://doi.org/10.1007/978-3-030-96087-2_8

We consider again the team of n unicycles

$$\begin{aligned} \dot{x}_i &= \begin{bmatrix} \cos(\theta_i) \\ \sin(\theta_i) \end{bmatrix} u_i, \\ \dot{\theta}_i &= \omega_i, \end{aligned} \tag{8.1}$$

with $i \in \mathbf{n}$, and recall from Chap. 3 that a formation is characterized, modulo roto-translations, via the formation vector $d \in \mathbb{R}^{2(n-1)}$. As in Chap. 7, we assume without loss of generality that the labeling of unicycles is chosen in such a way that unicycle 1 is at the front of the formation, i.e., $d_{1i}^1 \cdot e_1 \leq 0$ for all $i \in \mathbf{2}\!:\!\mathbf{n}$. The set of all formation vectors is

$$\mathsf{F} := \{d \in \mathbb{R}^{2(n-1)} : d_{1i}^1 \cdot e_1 \leq 0, \ i \in \mathbf{2}\!:\!\mathbf{n}\},$$

and the formation manifold is

$$\Gamma := \left\{(x, \theta) \in \mathbb{R}^{2n} \times \mathbb{T}^n : x_{1i} = R_1 d_{1i}^1, \ i \in \mathbf{n}\right\}.$$

8.2 Final Linear Motion

In this section, we present the solution to the following three formation problems with final linear motion.

Parallel Formation Flocking Problem (PFP)**.** Given a formation vector $d \in \mathsf{F}$ and a desired speed $\bar{v} > 0$, we want to design local and distributed feedbacks $(u_i^\star, \omega_i^\star)$, $i \in \mathbf{n}$, for the unicycle team in (8.1) that almost globally asymptotically stabilize the *parallel formation flocking manifold*

$$\Gamma_{pf} = \{(x, \theta) \in \Gamma : \theta_i = \theta_1, \ i \in \mathbf{2}\!:\!\mathbf{n}\},$$

with the constraint that $(u_i^\star, \omega_i^\star)|_{\Gamma_{pf}} = (\bar{v}, 0)$, $i \in \mathbf{n}$.

Parallel Formation Flocking Problem with a Beacon (PFP-B)**.** This problem is a variation of PFP when a beacon is available to all unicycles, i.e., when each unicycle can measure in its own body frame a common unit vector p with fixed angle $\theta_p \in \mathbb{S}^1$ measured with respect the inertial axis i_x of frame $\mathcal{I}$. In this case, we wish to design feedbacks $(u_i^\star, \omega_i^\star) : (y_i^i, \varphi_i, p^i) \mapsto (u_i, \omega_i)$, $i \in \mathbf{n}$, that almost globally asymptotically stabilize the *parallel formation flocking manifold with a beacon*

$$\Gamma_{pfb} := \left\{(x, \theta) \in \Gamma : \theta_i = \theta_p, \ i \in \mathbf{n}\right\},$$

with the constraint that $(u_i^\star, \omega_i^\star)|_{\Gamma_{pfb}} = (\bar{v}, 0)$, $i \in \mathbf{n}$.

Parallel Formation Path Following Problem (PPP). Given a formation vector $d \in \mathsf{F}$, a fixed straight line $\mathcal{C}(r_0, p)$, and a desired speed $\bar{v} > 0$, design feedbacks $(u_i^\star, \omega_i^\star) : (y_i^i, \varphi_i, p^i, \pi^i(x_i)) \mapsto (u_i, \omega_i)$, $i \in \mathbf{n}$ [where $\pi^i(x_i)$ is the minimum distance vector of x_i to the fixed straight line measured in body frame $\mathcal{B}_i$ defined in (3.7)] that almost globally asymptotically stabilize the *formation line path following manifold*

$$\Gamma_{lp} := \left\{(x, \theta) \in \Gamma_{pfb} : x_1 \in \mathcal{C}(r_0, p)\right\},$$

with the constraint that $(u_i^\star, \omega_i^\star)|_{\Gamma_{lp}} = (\bar{v}, 0)$, $i \in \mathbf{n}$.

Next, we develop ideas leading to the solution of these three problems. The focus for now is on developing the intuition for a plausible solution. Later, in Theorem 8.1, we confirm the theoretical validity of our arguments.

Let $d = (d_{i1}^1)_{i \in \mathbf{2}:\mathbf{n}} \in \mathsf{F}$ be the desired formation vector. Our point of departure is the insight developed in Chap. 7, where we attached to the origin of the generic unicycle i the offset vector $\delta_i = \alpha_i R_i e_1 + \beta_i R_i e_2$, where the constants α_i and β_i, defined in (7.3), are derived from the formation vector d. Letting $\hat{x}_i = x_i + \delta_i$ be the endpoint position of the offset vector in the coordinates of frame $\mathcal{I}$, we showed in Chap. 7 that synchronizing the endpoints $\hat{x}_i$ and the unicycles' heading angles θ_i is equivalent to stabilizing the desired formation with parallel headings (i.e., solving the Parallel Formation Problem, PP). For this reason, our discussion in what follows will concern the unicycle model (8.1) transformed in $(\hat{x}_i, \theta_i)_{i \in \mathbf{n}}$ coordinates,

$$\begin{aligned} \dot{\hat{x}}_i &= (u_i - \beta_i \omega_i) R_i e_1 + \alpha_i \omega_i R_i e_2, \\ \dot{\theta}_i &= \omega_i, \quad i \in \mathbf{n}. \end{aligned} \tag{8.2}$$

Recall the definition of y_i^i in Table 2.1 as the vector of relative positions x_{ij}^i available to robot i. In a similar fashion, we define

$$\hat{y}_i := (\hat{x}_{ij})_{j \in \mathcal{N}_i}, \quad \hat{y}_i^k := (\hat{x}_{ij}^k)_{j \in \mathcal{N}_i}. \tag{8.3}$$

Since $\hat{x}_{ij}^i = R_i^{-1}(x_{ij} + \delta_i - \delta_j) = x_{ij}^i + \alpha_i e_1 + \beta_i e_2 - \alpha_j R_j^i e_1 - \beta_j R_j^i e_2$ is a function of x_{ij}^i the constant vector $\alpha_i e_1 + \beta_i e_2$ and the relative orientation R_j^i, we see that $(\hat{y}_i)^i$ is a local and distributed quantity which can be used in the feedback of unicycle i.

Control design for PFP-B**.** In PFP-B, we want to stabilize the formation while synchronizing the heading angles to the angle θ_p of a beacon p, and imposing a steady-state linear motion with speed $\bar{v} > 0$. This problem looks very similar to PP, differing only in the common heading angle and steady-state speed requirements. To gain some intuition on how to address this problem, we revisit the control structure solving PP,

$$\begin{aligned} u_i &= u_i^\star(y_i^i, \varphi_i) = f_i(\hat{y}_i^i) \cdot e_1 + \beta_i \omega_i^\star(y_i^i, \varphi_i), \\ \omega_i &= \omega_i^\star(y_i^i, \varphi_i) = \frac{1}{\alpha_i}\left(f_i(\hat{y}_i^i) \cdot e_2 + k g_i(\varphi_i, \eta)\right), \ i \in \mathbf{n}. \end{aligned}$$

Substitution into (8.2) gives the closed-loop system

$$\begin{aligned} \dot{\hat{x}}_i &= f_i(\hat{y}_i) + k g_i(\varphi_i, \eta) R_i e_2, \\ \dot{\theta}_i &= \frac{1}{\alpha_i}\left(f_i(\hat{y}_i) \cdot R_i e_2 + k g_i(\varphi_i, \eta)\right), \quad i \in \mathbf{n}. \end{aligned}$$

With the goal in mind to adapt this controller to solve PFP-B, suppose we are able to change the unicycle controls in order to get the modified closed-loop dynamics

$$\begin{aligned} \dot{\hat{x}}_i &= f_i(\hat{y}_i) + \bar{v} p, \\ \dot{\theta}_i &= \frac{1}{\alpha_i}\left(f_i(\hat{y}_i) + \bar{v} p\right) \cdot R_i e_2, \quad i \in \mathbf{n}. \end{aligned} \tag{8.4}$$

Recall that $f_i(\hat{y}_i)$ in the $\dot{\hat{x}}_i$ equation is a single integrator consensus controller, and note that the term $\bar{v}p$ is a common bias vector independent of i which does not affect the convergence properties of the consensus controller. This implies that the endpoints $\hat{x}_i$, $i \in \mathbf{n}$, synchronize. As they do so, the signals $f_i(\hat{y}_i(t))$ in the $\dot{\theta}_i$ equation of (8.4) vanish as $t \to \infty$, so that the equation asymptotically reduces to

$$\dot{\theta}_i = \frac{1}{\alpha_i} \bar{v} p \cdot R_i e_2 = -\sin(\theta_i - \theta_p), \quad i \in \mathbf{n},$$

where we have used the fact that $p = \begin{bmatrix} \cos(\theta_p) & \sin(\theta_p) \end{bmatrix}^\top$. We quickly realize from the dynamics above that the unicycle headings θ_i almost globally synchronize to the common value θ_p, the angle of the beacon p. This reasoning of course needs to be made precise, but it points to the fact that if we were able to assign the closed-loop dynamics in (8.4), then we would expect to have solved PFP-B.[1]

Comparing the target closed-loop dynamics in (8.4) with (8.2), we deduce the following unicycle controls

$$\begin{aligned} u_i &= u_i^\star(y_i^i, \varphi_i) = f_i(\hat{y}_i^i) \cdot e_1 + \beta_i \omega_i^\star(y_i^i, \varphi_i) + \bar{v} p^i \cdot e_1, \\ \omega_i &= \omega_i^\star(y_i^i, \varphi_i) = \frac{1}{\alpha_i}\left(f_i(\hat{y}_i^i) \cdot e_2 + \bar{v} p^i \cdot e_2\right), \ i \in \mathbf{n}. \end{aligned} \tag{8.5}$$

Control design for PFP. The requirements of PFP are for the most part the same as those of PFP-B, but now the beacon is not available, and we want the unicycles' headings to synchronize to *some* common heading, as opposed to the beacon angle θ_p. This problem is harder than PFP-B, and we cannot attempt to achieve the target

[1] Provided some additional requirements in the statement of PFP-B are met.

dynamics similar to (8.4) because we no longer have a common inertial vector to be used as a bias of the consensus controller translating in a stabilizing effect for the heading angle dynamics of each unicycle.

The idea we propose to circumvent this problem is to assume the sensor digraph to be *hierarchical* in a sense made precise in Appendix B (see Definition B.1), but which can be informally described as follows. Suppose we may designate a leader who has no neighbors in the sensor digraph; then a set of first-layer followers that can only see the leader; then a set of second-layer followers that can see the leader and the first-layer followers, and so on. If this hierarchical sensor digraph assumption holds, then we may prescribe a linear motion for the leader and use the leader as a replacement for the beacon. Designating unicycle 1 as the leader, we could set $(u_1^\star, \omega_1^\star) = (\bar{v}, 0)$, making the leader move in a straight line in the direction of its initial heading and with speed $\bar{v}$. For the followers, suppose we are able to assign the following closed-loop dynamics in $(\hat{x}_i, \theta_i)_{i \in \mathbf{n}}$ coordinates,

$$\begin{aligned}\dot{\hat{x}}_i &= f_i(\hat{y}_i) + (\bar{v}/|\mathcal{N}_i|) \sum_{j \in \mathcal{N}_i} R_j e_1,\\ \dot{\theta}_i &= \frac{1}{\alpha_i}\left(f_i(\hat{y}_i) + (\bar{v}/|\mathcal{N}_i|) \sum_{j \in \mathcal{N}_i} R_j e_1 \right) \cdot R_i e_2, \quad i \in \mathbf{n}.\end{aligned} \tag{8.6}$$

For the first-layer followers, we have $\mathcal{N}_i = \{1\}$, and the above reduces to

$$\begin{aligned}\dot{\hat{x}}_i &= f_i(\hat{y}_i) + \bar{v} R_1 e_1,\\ \dot{\theta}_i &= \frac{1}{\alpha_i}\big(f_i(\hat{y}_i) + \bar{v} R_1 e_1\big) \cdot R_i e_2, \quad i \in \mathbf{n}.\end{aligned} \tag{8.7}$$

Since the leader moves along a straight line, the vector $R_1 e_1$ is constant and it can be regarded as a beacon. Accordingly, denoting $p := R_1 e_1$, we see the dynamics in (8.7) coincide with those in (8.4), which we have argued solve the problem, except that this time the unicycle headings will synchronize to the leader's heading, rather than the beacon angle. The arguments we presented so far suggest that the subgroup of unicycles given by the leader and first-layer followers will meet the specifications of PFP. Asymptotically, this entire group will constitute a beacon for the second-layer followers, and consequently, these will join the formation and synchronize their headings. Continuing this induction argument to cover the entire unicycle team, we expect that the target dynamics in (8.6) are a good candidate solution to PFP.

Comparing the target closed-loop dynamics in (8.6) with (8.2), we deduce the following unicycle controls

$$
\begin{aligned}
&(u_1^\star, \omega_1^\star) = (\bar{v}, 0), \\
&u_i^\star(y_i^i, \varphi_i, \mu_i^i) = f_i(\hat{y}_i^i) \cdot e_1 + \beta_i \omega_i^\star(y_i^i, \varphi_i, \mu_i^i) + \frac{\bar{v}}{|\mathcal{N}_i|} \sum_{j \in \mathcal{N}_i} R_j^i e_1 \cdot e_1, \\
&\omega_i^\star(y_i^i, \varphi_i, \mu_i^i) = \frac{1}{\alpha_i} \left(f_i(\hat{y}_i^i) \cdot e_2 + \frac{\bar{v}}{|\mathcal{N}_i|} \sum_{j \in \mathcal{N}_i} R_j^i e_1 \cdot e_2 \right), \quad i \in \mathbf{2{:}n}.
\end{aligned}
\tag{8.8}
$$

Control design for PPP. The control specifications of PPP are the same as those of PFP-B, with the additional requirement that the position x_1 of unicycle 1 converges to the fixed line $\mathcal{C}(r_0, p) = \{x \in \mathbb{R}^2 : x = r_0 + sp,\ s \in \mathbb{R}\}$, where p is a unit vector. Since in this problem we assume that unicycle i can measure the quantity p^i, effectively p is a beacon so it is natural to adopt as a starting point the controller developed to solve PFP-B and enhance it to make x_1 converge to $\mathcal{C}(r_0, p)$. The way we do this is to seek closed-loop dynamics in $(\hat{x}_i, \theta_i)_{i \in \mathbf{n}}$ coordinates which replace the beacon term by the single integrator path following controller $h(\cdot)$ defined in (4.7):

$$
\begin{aligned}
\dot{\hat{x}}_i &= f_i(\hat{y}_i) + h(\hat{x}_i), \\
\dot{\theta}_i &= \frac{1}{\alpha_i} \big(f_i(\hat{y}_i) + h(\hat{x}_i) \big) \cdot R_i e_2, \quad i \in \mathbf{n}.
\end{aligned}
$$

That these target dynamics make unicycle 1 converge to $\mathcal{C}(r_0, p)$ while preserving the remaining properties achieved in PFP-B is far from trivial, but we will show in the proof of Theorem 8.1 that indeed it does. The control input corresponding to the above target dynamics is

$$
\begin{aligned}
u_i &= u_i^\star(y_i^i, \varphi_i) = f_i(\hat{y}_i^i) \cdot e_1 + \beta_i \omega_i^\star(y_i^i, \varphi_i) + h(\hat{x}_i)^i \cdot e_1, \\
\omega_i &= \omega_i^\star(y_i^i, \varphi_i) = \frac{1}{\alpha_i} \big(f_i(\hat{y}_i^i) \cdot e_2 + h(\hat{x}_i)^i \cdot e_2 \big), \ i \in \mathbf{n}.
\end{aligned}
\tag{8.9}
$$

A unified controller structure. A comparison of the three controllers (8.5), (8.8), and (8.9) reveals a structural similarity. The three controllers have this form

$$
\begin{aligned}
u_i &= u_i^\star(y_i^i, \varphi_i, \mu_i^i) = f_i(\hat{y}_i^i) \cdot e_1 + \beta_i \omega_i^\star(y_i^i, \varphi_i, \mu_i^i) + \mu_i^i \cdot e_1, \\
\omega_i &= \omega_i^\star(y_i^i, \varphi_i, \mu_i^i) = \frac{1}{\alpha_i} \big(f_i(\hat{y}_i^i) \cdot e_2 + \mu_i^i \cdot e_2 \big), \ i \in \mathbf{n}.
\end{aligned}
\tag{8.10}
$$

where the functions μ_i^i are summarized in Table 8.1.

The next result confirms the theoretical validity of the foregoing informal considerations, and it states that the controller structure in (8.10) with auxiliary functions μ_i^i in Table 8.1 does indeed solve the three formation control problems with final linear motion.

Table 8.1 Auxiliary functions μ_i^i used to solve various formation problems with final linear motion

	Problem	Choice of function μ_i^i
A	PFP	$\mu_1^1 = \bar{v}e_1$, and $\mu_i^i = (\bar{v}/\lvert\mathcal{N}_i\rvert)\sum_{j\in\mathcal{N}_i} R_j^i e_1,\ i \in \mathbf{2:n}$
B	PFP-B	$\mu_i^i = \bar{v}p^i$
C	PPP	$\mu_i^i = h(\hat{x}_i)^i$, with $h(\cdot)$ defined in (4.7)

Theorem 8.1 (Solutions to PFP, PFP-B *and* PPP) *Consider the team of unicycles in* (8.1) *with directed sensor graph $\mathcal{G}$ containing a globally reachable node and any parameters $a_{ij} > 0$ for $i \in \mathbf{n}$, $j \in \mathcal{N}_i$. For each $d \in$ F, $\bar{v} > 0$, p such that $\|p\| = 1$, and $\bar{\alpha} > 0$, let $f_i(\cdot)$ be the single integrator consensus controller defined in* (4.2)*, let $h(\cdot)$ be the single integrator path following controller defined in* (4.7)*, and let α_i and β_i, $i \in \mathbf{n}$, be defined as in* (7.3)*. Letting $\hat{y}_i^i$ be defined as in* (8.3)*, the control inputs in* (8.10) *have the following properties:* (i) *(*PFP*) If $\mathcal{G}$ is a hierarchical digraph and μ_i^i is given as in row A of* Table 8.1*, then the control inputs in* (8.10) *are local and distributed, and they render the parallel formation flocking manifold Γ_{pf}* AGAS*, thus solving* PFP.

(ii) *(*PFP-B*) If μ_i^i is given as in row B of* Table 8.1*, then the control inputs in* (8.10) *are strictly functions of (y_i^i, φ_i, p^i), and they render the parallel formation flocking manifold Γ_{pfb}* AGAS*, thus solving* PFP-B.

(iii) *(*PPP*) If μ_i^i is given as in row C of* Table 8.1*, then the control inputs in* (8.10) *are strictly functions of $(y_i^i, \varphi_i, p^i, (c^\star(x_i) - x_i)^i)$, and they render the formation line path following manifold Γ_{lp}* AGAS*, thus solving CPP.*

The proof is in the next section.

Remark 8.1 As we pointed out earlier, for PPP the controller in (8.10) will enforce unicycle 1 to follow the line $C(r_0, p)$. If the line is to be followed by unicycle j as opposed to unicycle 1, one simply changes the value of β_i to $\beta_i = -d_{1i}^1 \cdot e_2 + d_{1j}^1 \cdot e_2$ for all $i \in \mathbf{n}$.

Remark 8.2 The control inputs in (8.10) for unicycle i are strictly functions of (y_i^i, φ_i), $(\alpha_j, \beta_j)_{j\in\mathcal{N}_i}$ and μ_i^i. In turn, in the case of PFP, μ_i^i in row A of Table 8.1 is also local and distributed. For PFP-B, μ_i^i in row B is a function of p^i. Finally, for PPP, μ_i^i in row C (with $k_0 = 1$ in $h(\cdot)$ for simplicity) satisfies

$$
\begin{aligned}
\mu_i^i = h(\hat{x}_i)^i &= (c^\star(\hat{x}_i) - \hat{x}_i)^i + \bar{v}p^i \\
&= (r_0 - \hat{x}_i)^i - ((r_0 - \hat{x}_i)^i \cdot p^i)p^i + \bar{v}p^i \\
&= (r_0 - x_i)^i - \alpha_i e_1 - \beta_i e_2 - ((r_0 - x_i)^i \cdot p^i)p^i + ((\alpha_i e_1 + \beta_i e_2) \cdot p^i)p^i + \bar{v}p^i \\
&= (c^\star(x_i) - x_i)^i - \alpha_i e_1 - \beta_i e_2 + ((\alpha_i e_1 + \beta_i e_2) \cdot p^i)p^i + \bar{v}p^i,
\end{aligned}
$$

which is a function of p^i and $(c^\star(x_i)-x_i)^i$, the displacement between x_i and its orthogonal projection $c^\star(x_i)$ on the path represented in body frame.

8.3 Proof of Theorem 8.1

Recall the unicycle model in (8.2) expressed in $(\hat{x}_i,\theta_i)_{i\in\mathbf{n}}$ coordinates

$$\begin{aligned}\dot{\hat{x}}_i &= (u_i-\beta_i\omega_i)R_ie_1+\alpha_i\omega_iR_ie_2,\\ \dot{\theta}_i &= \omega_i, \quad i\in\mathbf{n}.\end{aligned} \tag{8.11}$$

Since the dot product is invariant to a change of frame and $f_i(\hat{y}_i^i)=R_i^{-1}f_i(\hat{y}_i)$, it holds that $f_i(\hat{y}_i^i)\cdot e_1=R_if_i(\hat{y}_i^i)\cdot R_ie_1=f_i(\hat{y}_i)\cdot R_ie_1$. Similarly, $f_i(\hat{y}_i^i)\cdot e_2= f_i(\hat{y}_i)\cdot R_ie_2$ and $\mu_i^i\cdot e_1=\mu_i\cdot R_ie_1$. The control inputs in (8.10) represented with respect to the inertial frame therefore satisfy

$$\begin{aligned}u_i^\star &= f_i(\hat{y}_i)\cdot R_ie_1+\beta_i\omega_i^\star+\mu_i\cdot R_ie_1,\\ \omega_i^\star &= \frac{1}{\alpha_i}\left(f_i(\hat{y}_i)\cdot R_ie_2+\mu_i\cdot R_ie_2\right),\ i\in\mathbf{n}.\end{aligned} \tag{8.12}$$

Substituting $(u_i,\omega_i)=(u_i^\star,\omega_i^\star)$ in (8.12) into (8.11) and using the fact that $(f_i(\hat{y}_i)\cdot R_ie_1)R_ie_1+(f_i(\hat{y}_i)\cdot R_ie_2)R_ie_2=f_i(\hat{y}_i)$ and $(\mu_i\cdot R_ie_1)R_ie_1+(\mu_i\cdot R_ie_2)R_ie_2= \mu_i$ yields,

$$\begin{aligned}\dot{\hat{x}}_i &= f_i(\hat{y}_i)+\mu_i,\\ \dot{\theta}_i &= \frac{1}{\alpha_i}\left(f_i(\hat{y}_i)\cdot R_ie_2+\mu_i\cdot R_ie_2\right),\ i\in\mathbf{n}.\end{aligned} \tag{8.13}$$

Since the closed-loop system in (8.13) is globally Lipschitz for all choices of μ_i in Table 8.1, it has no finite escape times.

The target sets Γ_{pf}, Γ_{pfb}, and Γ_{lp} expressed in $(\hat{x},\theta)$ coordinates, denoted, respectively, by $\hat{\Gamma}_{pf}$, $\hat{\Gamma}_{pfb}$ and $\hat{\Gamma}_{lp}$, are given as follows

$$\hat{\Gamma}_{pf} := \left\{(\hat{x},\theta)\in\mathbb{R}^{2n}\times\mathbb{T}^n : \hat{x}_{1i}=0,\ \theta_{1i}=0,\ i\in\mathbf{2}{:}\mathbf{n}\right\}, \tag{8.14}$$

$$\hat{\Gamma}_{pfb} := \left\{(\hat{x},\theta)\in\hat{\Gamma}_{pf} : \theta_i=\theta_p,\ i\in\mathbf{n}\right\}, \tag{8.15}$$

$$\hat{\Gamma}_{lp} := \left\{(\hat{x},\theta)\in\hat{\Gamma}_{pfb} : \hat{x}_1\in C(r_0,p)\right\}. \tag{8.16}$$

We begin by proving the results for PFP-B and PPP corresponding to the sets $\hat{\Gamma}_{pfb}$ and $\hat{\Gamma}_{lp}$, respectively. As we will see, the proof of PFP-B is essentially the same as the proof of PPP just with some simplified arguments. We will then present the proof for PFP for hierarchical digraphs, corresponding to the set $\hat{\Gamma}_{pf}$.

Proof of PFP-B

Consider the closed-loop system (8.13) with $\mu_i = \bar{v}p$. It needs to be shown that the set $\hat{\Gamma}_{pfb}$ in (8.15) is AGAS.

Consider the new set of states $\hat{x}_1, \tilde{x} := (\hat{x}_{1i})_{i\in\mathbf{2:n}} \in \mathbb{R}^{2(n-1)}$ and $\theta = (\theta_i)_{i\in\mathbf{n}} \in \mathbb{T}^n$. The system equations in (8.13) can be expressed in terms of the new coordinates as

$$\begin{aligned}\dot{\hat{x}}_1 &= (f_1(\hat{y}_1) + \mu_1),\\ \dot{\hat{x}}_{1i} &= f_i(\hat{y}_i) - f_1(\hat{y}_1) + \mu_i - \mu_1,\ i \in \mathbf{2:n},\\ \dot{\theta}_i &= \frac{1}{\alpha_i}(f_i(\hat{y}_i)\cdot R_i e_2 + \mu_i \cdot R_i e_2),\ i \in \mathbf{n}.\end{aligned} \tag{8.17}$$

Note that $f_i(\hat{y}_i)$ is strictly a function of $\tilde{x}$, R_i is strictly a function of θ and $\mu_i = \bar{v}p$. Therefore, the right-hand side of (8.17) is independent of $\hat{x}_1$. The set $\hat{\Gamma}_{pfb}$ expressed in new coordinates is given by

$$\{(\hat{x}_1, \tilde{x}, \theta) \in \mathbb{R}^2 \times \mathbb{R}^{2(n-1)} \times \mathbb{T}^n : \tilde{x} = 0,\ \theta_i = \theta_p,\ i \in \mathbf{n}\},$$

where the angle of p in the inertial frame is denoted by θ_p. This set is also independent of $\hat{x}_1$. Therefore, since the closed-loop system has no finite escape times, the state $\hat{x}_1$ can be dropped. The remaining states are $(\tilde{x}, \theta) \in \mathbb{R}^{2(n-1)} \times \mathbb{T}^n$, and the set $\hat{\Gamma}_{pfb}$ reduces to the point

$$\mathcal{K} := \{(\tilde{x}, \theta) \in \mathbb{R}^{2(n-1)} \times \mathbb{T}^n : \tilde{x} = 0,\ \theta_i = \theta_p,\ i \in \mathbf{n}\},$$

and, since $\mu_i = \mu_1 = \bar{v}p$ for all $i \in \mathbf{2:n}$, system (8.17) reduces to

$$\begin{aligned}\dot{\hat{x}}_{1i} &= f_i(\hat{y}_i) - f_1(\hat{y}_1),\ i \in \mathbf{2:n},\\ \dot{\theta}_i &= \frac{1}{\alpha_i}(f_i(\hat{y}_i)\cdot R_i e_2 + \bar{v}p \cdot R_i e_2),\ i \in \mathbf{n}.\end{aligned} \tag{8.18}$$

Since $f_i(\hat{y}_i)$ is a single integrator consensus controller, it follows from (8.18) and the absence of finite escape times in the closed-loop system (8.13) that

$$\tilde{\Gamma}_2 := \{(\tilde{x}, \theta) \in \mathbb{R}^{2(n-1)} \times \mathbb{T}^n : \tilde{x} = 0\}$$

is globally asymptotically stable. This implies boundedness of $\tilde{x}$ and, in turn, boundedness of the states $(\tilde{x}, \theta)$. In the set $\tilde{\Gamma}_2$, the equations of motion of $(\tilde{x}, \theta)$ in (8.18) reduce to

$$\begin{aligned}\dot{\hat{x}}_{1i} &= 0,\ i \in \mathbf{2:n},\\ \dot{\theta}_i &= \frac{1}{\alpha_i}(\bar{v}p \cdot R_i e_2) = \frac{\bar{v}}{\alpha_i}\sin(\theta_p - \theta_i),\ i \in \mathbf{n},\end{aligned} \tag{8.19}$$

and the point $\mathcal{K}$ can be written as

$$\mathcal{K} = \{(\tilde{x}, \theta) \in \tilde{\Gamma}_2 : \theta_i = \theta_p,\ i \in \mathbf{n}\}.$$

To complete the proof, it needs to be shown that $\mathcal{K}$ is AGAS. In $\mathcal{K}$, the control inputs in (8.12) satisfy $u_i^\star = \bar{v}$ and $\omega_i^\star = 0$ for all $i \in \mathbf{n}$ as desired. Notice that the rotational system in (8.19) is decoupled for each unicycle $i \in \mathbf{n}$. As a consequence of Proposition 4.2, the set $\tilde{\Gamma}_1 = \mathcal{A} \cup \mathcal{K}$ is globally attractive relative to $\tilde{\Gamma}_2$, where $\mathcal{A}$ is the finite set of isolated equilibria given by

$$\mathcal{A} = \{(\tilde{x}, \theta) \in \tilde{\Gamma}_2 : \theta_i \in \{\theta_p, \theta_p + \pi\},\ i \in \mathbf{n}\} \backslash \mathcal{K}.$$

In the set $\mathcal{A}$, at least one unicycle has a heading angle 180° offset from the beacon while the remaining unicycle headings are aligned with the beacon. The set $\mathcal{K}$ is asymptotically stable relative to $\tilde{\Gamma}_2$, whereas the equilibria in $\mathcal{A}$ are exponentially unstable relative to $\tilde{\Gamma}_2$. It follows from Theorem A.6, setting $\Gamma_2 = \tilde{\Gamma}_2$ and $\Gamma_1 = \tilde{\Gamma}_1$, that $\mathcal{K}$ is AGAS. This proves part (ii) of the theorem.

Proof of PPP
Consider the closed-loop system (8.13) with $\mu_i = h(\hat{x}_i)$. It needs to be shown that the set $\hat{\Gamma}_{lp}$ in (8.16) is AGAS.

Recall that the path following controller in (4.7), $h(\hat{x}_i) = k_0(r_0 - \hat{x}_i) - k_0((r_0 - \hat{x}_i) \cdot p)p + \bar{v}p$, can be written as in (4.8)

$$h(\hat{x}_i) = k_0(c^\star(\hat{x}_i) - \hat{x}_i) + \bar{v}p, \tag{8.20}$$

where $(c^\star(\hat{x}_i) - \hat{x}_i) = (r_0 - \hat{x}_i) - ((r_0 - \hat{x}_i) \cdot p)p$ is the displacement between $\hat{x}_i$ and its orthogonal projection $c^\star(\hat{x}_i)$ on the path $\mathcal{C}(r_0, p)$. Let $p^\perp$ be a unit vector perpendicular to p, and consider the new states $\hat{x}_1^\parallel := \hat{x}_1 \cdot p$, $\hat{x}_1^\perp := (\hat{x}_1 - c^\star(\hat{x}_1)) \cdot p^\perp = (\hat{x}_1 - r_0) \cdot p^\perp$, $\tilde{x} := (\hat{x}_{1i})_{i \in \mathbf{2:n}} \in \mathbb{R}^{2(n-1)}$ and $\theta = (\theta_i)_{i \in \mathbf{n}} \in \mathbb{T}^n$ defined under the diffeomorphism

$$\begin{aligned} &F : \mathbb{R}^{2n} \times \mathbb{T}^n \to \mathbb{R} \times \mathbb{R} \times \mathbb{R}^{2(n-1)} \times \mathbb{T}^n, \\ &F(\hat{x}, \theta) = (\hat{x}_1^\parallel, \hat{x}_1^\perp, \tilde{x}, \theta), \end{aligned}$$

with smooth inverse yielding

$$\begin{aligned} \hat{x}_1 &= \hat{x}_1^\parallel p + (\hat{x}_1^\perp + r_0 \cdot p^\perp)p^\perp, \\ (\hat{x}_i)_{i \in \mathbf{2:n}} &= (\hat{x}_{1i} + (\hat{x}_1^\parallel p + (\hat{x}_1^\perp + r_0 \cdot p^\perp)p^\perp))_{i \in \mathbf{2:n}}, \\ \theta &= (\theta_i)_{i \in \mathbf{n}}. \end{aligned}$$

The system equations in (8.13) can be expressed in terms of new coordinates as

$$\begin{aligned}
\dot{\hat{x}}_1^{\parallel} &= (f_1(\hat{y}_1) + \mu_1) \cdot p, \\
\dot{\hat{x}}_1^{\perp} &= (f_1(\hat{y}_1) + \mu_1) \cdot p^{\perp} - \dot{c}^{\star}(\hat{x}_1) \cdot p^{\perp}, \\
\dot{\hat{x}}_{1i} &= f_i(\hat{y}_i) - f_1(\hat{y}_1) + \mu_i - \mu_1, \ i \in \mathbf{2}\!:\!\mathbf{n}, \\
\dot{\theta}_i &= \frac{1}{\alpha_i}(f_i(\hat{y}_i) \cdot R_i e_2 + \mu_i \cdot R_i e_2), \ i \in \mathbf{n}.
\end{aligned} \tag{8.21}$$

Next it is shown that (8.21) is a dynamic system whose right-hand side is independent of the state $\hat{x}_1^{\parallel}$. To see this note that $f_i(\hat{y}_i) = \sum_{j \in \mathcal{N}_i} a_{ij}(\hat{x}_{1j} - \hat{x}_{1i})$ is strictly a function of $\tilde{x}$, R_i is strictly a function of θ. Moreover, using the fact that

$$\begin{aligned}
c^{\star}(\hat{x}_i) - \hat{x}_i &= (r_0 - \hat{x}_i) - ((r_0 - \hat{x}_i) \cdot p)p, \\
&= (r_0 - \hat{x}_1) - ((r_0 - \hat{x}_1) \cdot p)p - \hat{x}_{1i} + (\hat{x}_{1i} \cdot p)p, \\
&= (c^{\star}(\hat{x}_1) - \hat{x}_1) - \hat{x}_{1i} + (\hat{x}_{1i} \cdot p)p,
\end{aligned}$$

and that $c^{\star}(\hat{x}_1) - \hat{x}_1$ is parallel to $p^{\perp}$, it follows that

$$\begin{aligned}
\mu_i = h(\hat{x}_i) &= -k_0[(\hat{x}_1 - c^{\star}(\hat{x}_1)) \cdot p^{\perp}]p^{\perp} - k_0\hat{x}_{1i} + k_0(\hat{x}_{1i} \cdot p)p + \bar{v}p \\
&= -k_0\hat{x}_1^{\perp}p^{\perp} - k_0\hat{x}_{1i} + k_0(\hat{x}_{1i} \cdot p)p + \bar{v}p,
\end{aligned}$$

which is strictly a function of $\hat{x}_1^{\perp}$ and $\tilde{x}$. Since $c^{\star}(\hat{x}_1)$ lies on the line perpendicular to $p^{\perp}$ at all times, it follows that $\dot{c}^{\star}(\hat{x}_1) \cdot p^{\perp} = 0$. Therefore,

$$\dot{\hat{x}}_1^{\perp} = (f_1(\hat{y}_1) + \mu_1) \cdot p^{\perp}$$

and (8.21) is a dynamical system whose right-hand side is independent of $\hat{x}_1^{\parallel}$. The set $\hat{\Gamma}_{lp}$ expressed in new coordinates is

$$\{(\hat{x}_1^{\parallel}, \hat{x}_1^{\perp}, \tilde{x}, \theta) \in \mathbb{R} \times \mathbb{R} \times \mathbb{R}^{2(n-1)} \times \mathbb{T}^n : \hat{x}_1^{\perp} = 0, \ \tilde{x} = 0, \ \theta_i = \theta_p, \ i \in \mathbf{n}\},$$

where the angle of p in the inertial frame is denoted by θ_p. This set is also independent of $\hat{x}_1^{\parallel}$. Therefore since there are no finite escape times, the state $\hat{x}_1^{\parallel}$ can be dropped. The remaining states are $(\hat{x}_1^{\perp}, \tilde{x}, \theta) \in \mathbb{R} \times \mathbb{R}^{2(n-1)} \times \mathbb{T}^n$ and the set $\hat{\Gamma}_{lp}$ reduces to the point

$$\mathcal{K} := \{(\hat{x}_1^{\perp}, \tilde{x}, \theta) \in \mathbb{R} \times \mathbb{R}^{2(n-1)} \times \mathbb{T}^n : \hat{x}_1^{\perp} = 0, \ \tilde{x} = 0, \ \theta_i = \theta_p, \ i \in \mathbf{n}\}.$$

Since $c^{\star}(\hat{x}_i) - \hat{x}_i$ is parallel to $p^{\perp}$, it holds that $\mu_i = [k_0(c^{\star}(\hat{x}_i) - \hat{x}_i) \cdot p^{\perp}]p^{\perp} + \bar{v}p$ for $i \in \mathbf{n}$. Moreover, since $c^{\star}(\hat{x}_i)$ lies on $\mathcal{C}(r_0, p)$ for all $i \in \mathbf{n}$, it follows that $(c^{\star}(\hat{x}_i) - c^{\star}(\hat{x}_1)) \cdot p^{\perp} = 0$ for all $i \in \mathbf{2}\!:\!\mathbf{n}$. These two facts imply that $\mu_i - \mu_1 = [k_0(c^{\star}(\hat{x}_i) - c^{\star}(\hat{x}_1)) \cdot p^{\perp} - k_0(\hat{x}_i - \hat{x}_1) \cdot p^{\perp}]p^{\perp} = (k_0\hat{x}_{i1} \cdot p^{\perp})p^{\perp}$ in (8.21) for all $i \in \mathbf{2}\!:\!\mathbf{n}$. Therefore, system (8.21) can be written as

$$
\begin{aligned}
\dot{\hat{x}}_1^{\perp} &= (f_1(\hat{y}_1) + \mu_1) \cdot p^{\perp}, \\
\dot{\hat{x}}_{1i} \cdot p &= (f_i(\hat{y}_i) - f_1(\hat{y}_1)) \cdot p, \ i \in \mathbf{2}\colon\mathbf{n}, \\
\dot{\hat{x}}_{1i} \cdot p^{\perp} &= (f_i(\hat{y}_i) - f_1(\hat{y}_1)) \cdot p^{\perp} + k_0 \hat{x}_{i1} \cdot p^{\perp} \\
&= (\hat{f}_i(\hat{y}_i, \hat{x}_{i1}) - \hat{f}_1(\hat{y}_1, 0)) \cdot p^{\perp}, \ i \in \mathbf{2}\colon\mathbf{n}, \\
\dot{\theta}_i &= \frac{1}{\alpha_i}(f_i(\hat{y}_i) \cdot R_i e_2 + \mu_i \cdot R_i e_2), \ i \in \mathbf{n},
\end{aligned}
\tag{8.22}
$$

where the state $\tilde{x}$ has been split into components parallel $(\hat{x}_{1i} \cdot p)$ and perpendicular $(\hat{x}_{1i} \cdot p^{\perp})$ to p and $\hat{f}_i(\hat{y}_i, \hat{x}_{i1}) := f_i(\hat{y}_i) + k_0 \hat{x}_{i1}$ represents the consensus control law $f_i(\hat{y}_i)$ with an additional edge added from unicycle i to unicycle 1. Therefore, both $f_i(\hat{y}_i)$ and $\hat{f}_i(\hat{y}_i, \hat{x}_{i1})$ are integrator consensus control laws. In (8.22), $f_i(\hat{y}_i) \cdot p = \sum_{j \in \mathcal{N}_i} a_{ij}(\hat{x}_{1j} \cdot p - \hat{x}_{1i} \cdot p)$ is strictly a function of the elements of $\tilde{x}$ in the direction of p and $f_i(\hat{y}_i) \cdot p^{\perp} = \sum_{j \in \mathcal{N}_i} a_{ij}(\hat{x}_{1j} \cdot p^{\perp} - \hat{x}_{1i} \cdot p^{\perp})$ is strictly a function of the elements of $\tilde{x}$ in the direction of $p^{\perp}$. It follows from the equations of motion for path following in (8.22) and the absence of finite escape times in the closed-loop system (8.13) that

$$
\tilde{\Gamma}_3 := \{(\hat{x}_1^{\perp}, \tilde{x}, \theta) \in \mathbb{R}^2 \times \mathbb{R}^{2(n-1)} \times \mathbb{T}^n : \tilde{x} = 0\}
$$

is globally asymptotically stable. This implies boundedness of $\tilde{x}$ and, in turn, boundedness of $f_i(\hat{y}_i)$ for all $i \in \mathbf{n}$. Therefore $\hat{x}_1^{\perp}$ satisfies

$$
\dot{\hat{x}}_1^{\perp} = f_1(\hat{y}_1) \cdot p^{\perp} + \mu_1 \cdot p^{\perp} \leq B - k_0 \hat{x}_1^{\perp},
$$

where $B = \sup \|f_1(\hat{y}_1)\| < \infty$. Since the origin of the nominal system $\dot{\hat{x}}_1^{\perp} = -k_0 \hat{x}_1^{\perp}$ is GAS, this implies boundedness of $\hat{x}_1^{\perp}$ and, in turn, boundedness of the states $(\hat{x}_1^{\perp}, \tilde{x}, \theta)$. In the set $\tilde{\Gamma}_3$, $f_i(\hat{y}_i) = 0$ for all $i \in \mathbf{n}$ and the equations of motion in (8.22) reduce to

$$
\begin{aligned}
\dot{\hat{x}}_1^{\perp} &= \mu_1 \cdot p^{\perp} = -k_0 \hat{x}_1^{\perp}, \\
\dot{\hat{x}}_{1i} &= 0, \ i \in \mathbf{2}\colon\mathbf{n}, \\
\dot{\theta}_i &= \frac{1}{\alpha_i}(\mu_i \cdot R_i e_2), \ i \in \mathbf{n}.
\end{aligned}
\tag{8.23}
$$

For system (8.23), $\hat{x}_1^{\perp} \to 0$ as $t \to \infty$ and the compact set

$$
\tilde{\Gamma}_2 := \{(\hat{x}_1^{\perp}, \tilde{x}, \theta) \in \tilde{\Gamma}_3 : \hat{x}_1^{\perp} = 0, \ i \in \mathbf{n}\},
$$

diffeomorphic to $\mathbb{T}^n$, is an embedded submanifold of $\mathbb{R}^2 \times \mathbb{R}^{2(n-1)} \times \mathbb{T}^n$ and is globally asymptotically stable relative to $\tilde{\Gamma}_3$. By Theorem A.5, $\tilde{\Gamma}_2$ is globally asymptotically stable. In the set $\tilde{\Gamma}_2$, $\hat{x}_i = c^{\star}(\hat{x}_i)$ and $\mu_i = \bar{v}p$ for all $i \in \mathbf{n}$ and the equations of motion of $(\hat{x}_1^{\perp}, \tilde{x}, \theta)$ in (8.23) reduce to

$$\begin{aligned} \dot{\hat{x}}_1^\perp &= 0, \\ \dot{\hat{x}}_{1i} &= 0, \ i \in \mathbf{2}:\mathbf{n}, \\ \dot{\theta}_i &= \frac{1}{\alpha_i}(\bar{v}p \cdot R_i e_2) = \frac{\bar{v}}{\alpha_i}\sin(\theta_p - \theta_i), \ i \in \mathbf{n}. \end{aligned} \tag{8.24}$$

The point $\mathcal{K}$ can be written as

$$\mathcal{K} = \{(\hat{x}_1^\perp, \tilde{x}, \theta) \in \tilde{\Gamma}_2 : \theta_i = \theta_p, \ i \in \mathbf{n}\}.$$

To complete the proof, it needs to be shown that $\mathcal{K}$ is AGAS. This follows using the same arguments as in the proof of PFP-B. This proves part (iii) of the theorem.

Proof of PFP

Now, we present the proof for PFP for hierarchical digraphs. Consider the closed-loop system (8.13) with $\mu_1 = R_1 e_1$ and $\mu_i = (\bar{v}/|\mathcal{N}_i|)\sum_{j\in\mathcal{N}_i} R_j e_1$ for $i \in \mathbf{2}:\mathbf{n}$. The leader's control inputs represented with respect to the inertial frame are given by

$$u_1^\star = \bar{v}, \ \omega_1^\star = 0, \tag{8.25}$$

and the control inputs of the follower unicycles are given by

$$\begin{aligned} u_i^\star &= f_i(\hat{y}_i) \cdot R_i e_1 + \beta_i \omega_i^\star + \frac{\bar{v}}{|\mathcal{N}_i|}\sum_{j\in\mathcal{N}_i} R_j e_1 \cdot R_i e_1, \\ \omega_i^\star &= \frac{1}{\alpha_i}\left(f_i(\hat{y}_i) \cdot R_i e_2 + \frac{\bar{v}}{|\mathcal{N}_i|}\sum_{j\in\mathcal{N}_i} R_j e_1 \cdot R_i e_2 \right), \end{aligned} \tag{8.26}$$

where $\sum_{j\in\mathcal{N}_i} R_j e_1 \cdot R_i e_1 = \sum_{j\in\mathcal{N}_i}\cos(\theta_{ij})$ and $\sum_{j\in\mathcal{N}_i} R_j e_1 \cdot R_i e_2 = \sum_{j\in\mathcal{N}_i}\sin(\theta_{ij})$.

It needs to be shown that $\hat{\Gamma}_{pf}$ is AGAS. The closed-loop equation for the leader is $\dot{\hat{x}}_1 = \bar{v}R_1 e_1, \ \dot{\theta}_1 = 0$. This implies that the leader moves in a straight line at the desired speed $\bar{v}$ in the direction of its initial heading. In this case, since the heading vector $p := R_1 e_1$ is a constant vector with angle $\theta_p := \theta_1$, we can view it as a beacon.

Define the vertex set $\mathcal{L}_j \subset \mathcal{V}$ to be the set of unicycles in layer j of the hierarchical digraph and $\bar{\mathcal{L}}_j := \cup_{i=1}^j \mathcal{L}_i$. The equations of motion of unicycles in the set $\bar{\mathcal{L}}_j$ are independent of the states of unicycles outside of this set. In the development that follows, we will consider relative translational and absolute rotational coordinates of the unicycles in the isolated set $\bar{\mathcal{L}}_j$. These states are denoted by $(\tilde{x}, \theta)_{\bar{\mathcal{L}}_j} := ((\hat{x}_{1i})_{i\in\bar{\mathcal{L}}_j\setminus\{1\}}, (\theta_i)_{i\in\bar{\mathcal{L}}_j}) \in \mathsf{X}_j$. In the reduced state space X_j, the system dynamics are given by

$$\begin{aligned}\dot{\hat{x}}_{1i} &= f_i + \frac{\bar{v}}{|\mathcal{N}_i|}\sum_{j\in\mathcal{N}_i} R_j e_1 - \bar{v}p,\ i \in \bar{\mathcal{L}}_j\backslash\{1\},\\ \dot{\theta}_i &= \omega_i,\quad i \in \bar{\mathcal{L}}_j,\end{aligned} \tag{8.27}$$

where $f_i(\hat{y}_i)$ and R_i are strictly functions of $(\tilde{x},\theta)_{\bar{\mathcal{L}}_j}$. Therefore, system (8.27) defines an autonomous dynamical system in the state space X_j. Correspondingly, define the reduced formation flocking manifold $\tilde{\Gamma}_{\bar{\mathcal{L}}_j} = \{(\tilde{x},\theta)_{\bar{\mathcal{L}}_j} \in \mathsf{X}_j : \hat{x}_{1i} = 0,\ \theta_i = \theta_p,\ i \in \bar{\mathcal{L}}_j\}$. In the set $\tilde{\Gamma}_{\bar{\mathcal{L}}_j}$, $f_i(\hat{y}_i) = 0$ and $p = R_i e_1$, and therefore, the control inputs in (8.25)–(8.26) satisfy $u_i^\star = \bar{v}$ and $\omega_i^\star = 0$ for all $i \in \bar{\mathcal{L}}_j$.

Letting m be the total number of layers in the hierarchical digraph, the set $\hat{\Gamma}_{pf}$ expressed in coordinates $(\tilde{x},\theta)_{\bar{\mathcal{L}}_m} \in \mathsf{X}_m = \mathbb{R}^{2(n-1)} \times \mathbb{T}^n$ becomes

$$\tilde{\Gamma}_{pf} := \tilde{\Gamma}_{\bar{\mathcal{L}}_m} = \left\{(\tilde{x},\theta)_{\bar{\mathcal{L}}_m} \in \mathbb{R}^{2(n-1)} \times \mathbb{T}^n : \hat{x}_{1i} = 0,\ \theta_i = \theta_p,\ i \in \bar{\mathcal{L}}_m\right\}.$$

In order to solve the formation control problem, it needs to be shown that $\tilde{\Gamma}_{pf}$ is AGAS. An induction approach based on reduction will be employed. Consider first the isolated set $\bar{\mathcal{L}}_2$. The leader unicycle with heading vector $p = R_1 e_1$ is the unique neighbor for all follower unicycles in $\mathcal{L}_2$ whose control inputs in (8.26) reduce to

$$\begin{aligned}u_i^\star &= f_i(\hat{y}_i)\cdot R_i e_1 + \beta_i \omega_i^\star + \bar{v}p \cdot R_i e_1,\\ \omega_i^\star &= \frac{1}{\alpha_i}\left(f_i(\hat{y}_i)\cdot R_i e_2 + \bar{v}p\cdot R_i e_2\right),\ i \in \mathcal{L}_2.\end{aligned} \tag{8.28}$$

This has the same form as in PFP-B, and by the same arguments as in the proof of PFP-B, the set $\tilde{\Gamma}_{\bar{\mathcal{L}}_2}$ is AGAS for the state space X_2. Now for $k \in \mathbb{N}$, consider the isolated node set $\bar{\mathcal{L}}_{k-1}$ and assume the set $\tilde{\Gamma}_{\bar{\mathcal{L}}_{k-1}}$ is AGAS in X_{k-1} coordinates with domain of attraction $\mathsf{X}_{k-1}\backslash N_{k-1}$ where N_{k-1} is a set of Lebesgue measure zero. Next it will be shown that this implies that $\Gamma_{\bar{\mathcal{L}}_k}$ is AGAS for the isolated system in X_k coordinates. Let $\tilde{\Gamma}_3$ denote the insertion of $\tilde{\Gamma}_{\bar{\mathcal{L}}_{k-1}}$ into the state space X_k, i.e.,

$$\tilde{\Gamma}_3 := \{(\tilde{x},\theta)_{\bar{\mathcal{L}}_k} \in \mathsf{X}_k : \hat{x}_{1i} = 0,\ \theta_i = \theta_p,\ i \in \bar{\mathcal{L}}_{k-1}\}.$$

The set $\tilde{\Gamma}_3$ is AGAS because the closed-loop system in (8.13) has no finite escape times. The domain of attraction of $\tilde{\Gamma}_3$ is given by $\mathsf{X}_k\backslash N$, where

$$N := \{(\tilde{x},\theta)_{\bar{\mathcal{L}}_k} \in \mathsf{X}_k : (\tilde{x},\theta)_{\bar{\mathcal{L}}_{k-1}} \in N_{k-1}\}$$

is the insertion of N_{k-1} into X_k and is a set of Lebesgue measure zero. The unicycles in the set $\mathcal{L}_k$ have neighbors strictly in $\bar{\mathcal{L}}_{k-1}$, and their equations of motion are given by

$$
\begin{aligned}
\dot{\hat{x}}_{1i} &= (f_i - f_1) + \frac{\bar{v}}{|\mathcal{N}_i|} \sum_{j \in \mathcal{N}_i} R_j e_1 - \bar{v} p, \\
\dot{\theta}_i &= \omega_i, \quad i \in \mathcal{L}_k,
\end{aligned} \tag{8.29}
$$

where $f_1(\hat{y}_1) = 0$ represents the consensus controller for the leader without neighbors. The term $(\bar{v}/|\mathcal{N}_i|) \sum_{j \in \mathcal{N}_i} R_j e_1 - \bar{v} p$ is bounded since it is a continuous function of the bounded quantities $(\theta_i)_{\bar{\mathcal{L}}_{k-1}}$ and it vanishes in the set $\tilde{\Gamma}_3$ because $R_j e_1 = p$ for all $j \in \bar{\mathcal{L}}_{k-1}$. For all $i \in \mathcal{L}_k \backslash \{1\}$, $\hat{x}_{1i}$ is bounded because the origin is exponentially stable for the nominal linear system

$$
\dot{\hat{x}}_{1i} = (f_i - f_1), \; i \in \mathcal{L}_k.
$$

Therefore, all closed-loop solutions in X_k remain bounded. In the set $\tilde{\Gamma}_3$, the equations of motion for unicycles in $\mathcal{L}_k$ in (8.29) simplify to

$$
\begin{aligned}
\dot{\hat{x}}_{1i} &= f_i(\hat{y}_i) - f_1(\hat{y}_1), \\
\dot{\theta}_i &= \frac{1}{\alpha_i} \left(f_i(\hat{y}_i) \cdot R_i e_2 + \bar{v} p \cdot R_i e_2 \right), \quad i \in \mathcal{L}_k.
\end{aligned} \tag{8.30}
$$

It follows that the compact set

$$
\tilde{\Gamma}_2 := \{(\tilde{x}, \theta)_{\bar{\mathcal{L}}_k} \in \tilde{\Gamma}_3 : \hat{x}_{1i} = 0, \; i \in \mathcal{L}_k\}
$$

is GAS relative to $\tilde{\Gamma}_3$ and reduction in Theorem A.5 implies that $\tilde{\Gamma}_2$ is GAS relative to $\mathsf{X}_k \backslash N$, a positively invariant set of full measure. We conclude that $\tilde{\Gamma}_2$ is AGAS in X_k. In the set $\tilde{\Gamma}_2$, the rotational dynamics satisfy $\dot{\theta}_i = (\bar{v}/\alpha_i)\sin(\theta_p - \theta_i)$ for $i \in \mathcal{L}_k$ and the point

$$
\mathcal{K} := \tilde{\Gamma}_{\bar{\mathcal{L}}_k} = \{(\tilde{x}, \theta)_{\bar{\mathcal{L}}_k} \in \tilde{\Gamma}_2 : \theta_i = \theta_p, \; i \in \mathcal{L}_k\}
$$

is AGAS relative to $\tilde{\Gamma}_2$ with an additional finite number of exponentially unstable isolated equilibria

$$
\mathcal{A} = \{(\tilde{x}, \theta)_{\bar{\mathcal{L}}_k} \in \tilde{\Gamma}_2 : \theta_i \in \{\theta_p, \theta_p + \pi\}, \; i \in \mathcal{L}_k\} \backslash \mathcal{K}.
$$

The set $\tilde{\Gamma}_1 = \mathcal{A} \cup \mathcal{K}$ is globally attractive with respect to $\tilde{\Gamma}_2$ and applying Theorem A.6, setting $\Gamma_2 = \tilde{\Gamma}_2$ and $\Gamma_1 = \tilde{\Gamma}_1$, the point $\mathcal{K}$ is AGAS relative to $\mathsf{X}_k \backslash N$ with domain of attraction $(\mathsf{X}_k \backslash N) \backslash M$ where M is a set of Lebesgue measure. Equivalently, $\mathcal{K}$ is AGAS relative to X_k with domain of attraction $\mathsf{X}_k \backslash N_k$ where $N_k = N \cup M$ is a set of Lebesgue measure zero. It follows by induction that $\tilde{\Gamma}_{pf}$ is AGAS. This proves part (i) of the theorem. □

8.4 Final Circular Motion

In this section, we present the solution to the following two formation problems with final circular motion.

Circular Formation Flocking Problem (CFP). Given a formation vector $d \in \mathsf{F}$, a radius of rotation $\beta_1 > 0$, and a desired speed $\bar{v} > 0$, design local and distributed feedbacks $(u_i^\star, \omega_i^\star)$, $i \in \mathbf{n}$, that almost globally asymptotically stabilize the *circular formation flocking manifold*

$$\Gamma_{cf} := \left\{(x, \theta) \in \mathbb{R}^{2n} \times \mathbb{T}^n : \hat{x}_i = \hat{x}_1, \theta_i - \theta_1 = \rho_i(d, \beta_1),\ i \in \mathbf{2}\!:\!\mathbf{n}\right\},$$

under the constraint that $(u_i^\star, \omega_i^\star)|_{\Gamma_{cf}} = (\bar{v}\beta_i, \bar{v})$.

We recall from Sect. 3.3.4 that in the definition of Γ_{cf}, we have that:

- $\beta_i := \sqrt{(d_{1i}^1 \cdot e_1)^2 + (\beta_1 - d_{1i}^1 \cdot e_2)^2}$ is the radius of the circle that the ith unicycle must traverse to stay in formation.
- The vector $\hat{x}_i := x_i + \beta_i R_i e_2$ is the center of the circle of radius β_i that unicycle i would trace if its translational speed were $\bar{v}\beta_i$ and its angular speed where $\bar{v}$.
- The function $\rho_i(d, \beta_1) := \text{atan2}(d_{1i}^1 \cdot e_1, \beta_1 - d_{i1}^1 \cdot e_2)$ represents the desired offset between the heading of unicycle i and that of unicycle 1. We let $\rho_{ij}(d, \beta_1) := \rho_j(d, \beta_1) - \rho_i(d, \beta_1)$ and $\rho(d, \beta_1) := (\rho_i(d, \beta_1))_{i \in \mathbf{2}:\mathbf{n}}$.

As we did in (8.3), in what follows we denote $\hat{y}_i := (\hat{x}_{ij})_{j \in \mathcal{N}_i}$, and $\hat{y}_i^k := (\hat{x}_{ij}^k)_{j \in \mathcal{N}_i}$.

Circular Formation Path Following Problem (CPP). This problem is similar to CFP, but now the center of the circle which we want the formation to approach is a given vector $c \in \mathbb{R}^2$ fixed a priori, and the target set is the *formation circle path following manifold*

$$\Gamma_{cp} := \left\{(x, \theta) \in \Gamma_{cf} : \hat{x}_1 = c\right\},$$

which we want to almost globally asymptotically stabilize by means of feedbacks $(u_i^\star, \omega_i^\star) : (y_i^i, \varphi_i, (c - x_i)^i) \mapsto (u_i, \omega_i)$, $i \in \mathbf{n}$, under the constraint that $(u_i^\star, \omega_i^\star)|_{\Gamma_{cp}} = (\bar{v}\beta_i, \bar{v})$.

A special case of CFP: formations on a common circle. Our point of departure in solving CFP is the solution, due to [1], of the special case when the unicycles in formation lie on a common circle. In our setup, this corresponds to choosing $\beta_1 = \cdots = \beta_n$. Letting $r = \beta_1$ be the radius of the common circle (and consequently $\hat{x}_i = x_i + r R_i e_2$), [1] proposed the following feedbacks

$$\begin{aligned} u_i &= u_i^\star(\varphi_i) = r\bar{v} + k g_i((\theta_{ij} - \rho_{ij}(d, r))_{j \in \mathcal{N}_i}), \\ \omega_i &= \omega_i^\star(\varphi_i, \nu_i(y_i^i)) = \frac{u_i^\star(\varphi_i)}{r} + K\hat{\varphi}(\nu_i(\hat{y}_i^i))\nu_i(\hat{y}_i^i),\ i \in \mathbf{n}, \end{aligned} \tag{8.31}$$

where:

- $k, K > 0$ are design parameters;
- $g_i(\cdot)$ is the Kuramoto consensus controller presented in (4.14);
- $\hat{\varphi}(v_i) := 1/(1 + \|v_i\|)$ is a decentralized saturation function; and
- $v_i(\hat{y}_i^i) = -f(\hat{y}_i^i) \cdot e_1$, and $f(\cdot)$ is the single integrator consensus controller in (4.2).

In [1] it is shown that if the sensor graph is connected and undirected, then for sufficiently small k and K the feedbacks (8.31) render the circular formation flocking manifold Γ_{cf} asymptotically stable.

Control design for CFP. In order to solve CFP, we need to enhance the controller (8.31) in two directions. First, we need to render the controller an *almost global* stabilizer. Second, we need to adapt the controller to allow for unequal values of β_i.

To address the first problem, we replace the Kuramoto consensus controller $g_i(\cdot)$ in (8.31) by its almost global stabilizing counterpart reviewed in Chap. 4, the rotational integrator consensus controller $g_i((\theta_{ij} - \rho_{ij}(d, r))_{j\in\mathcal{N}_i}, \mathbf{1})$ in[2] (4.15), with control parameters b_{ij}. Using Theorem 4.2 [which asserts the almost global synchronization of rotational integrators under the feedback in (4.15)] and Theorem A.6 (the novel reduction theorem for almost global asymptotic stability of sets), we will show in Theorem 8.2 that, with this modification, for almost all values of $b_{ij} > 0$ in $g_i(\cdot, \cdot)$, the controller (8.31) almost globally asymptotically stabilizes Γ_{cf} in the special case $r = \beta_1 = \cdots = \beta_n$.

Next, we need to further enhance the controller to allow for unequal values of β_i, $i \in \mathbf{n}$. For this, we fix $r > 0$ and define $\gamma_i := \beta_i - r$ and $z_i := x_i + \gamma_i R_i e_2$. Consider further the feedback transformation $u_i = \hat{u}_i + \gamma_i \omega_i$. In (z, θ) coordinates and with the given feedback transformation, system (8.1) reads as

$$\begin{aligned} \dot{z}_i &= u_i R_i e_1 - \gamma_i \omega_i R_i e_1 = \hat{u}_i R_i e_1, \\ \dot{\theta}_i &= \omega_i, \end{aligned} \tag{8.32}$$

where $(\hat{u}_i, \omega_i)$ is the new control input. System (8.32) represents a unicycle with position z_i, heading θ_i, and control input $(\hat{u}_i, \omega_i)$. This unicycle has parallel heading to the unicycle with state (x_i, θ_i), but its position is offset by the constant distance $\gamma_i = \beta_i - r$ along its second body axis, b_{iy}. In what follows, we will refer to the model (8.32) for $i \in \mathbf{n}$ as the *virtual unicycle team.*

Define the offset position $\hat{z}_i := z_i + r R_i e_2$ and the set

$$\hat{\Gamma}_{cf} = \{(\hat{z}, \theta) \in \mathbb{R}^{2n} \times \mathbb{T}^n : \hat{z}_i = \hat{z}_1,\ \theta_i - \theta_1 = \rho_i(d, \beta_1),\ i \in \mathbf{2}:\mathbf{n}\}. \tag{8.33}$$

Using the fact that $z_i = x_i + \gamma_i R_i e_2$, that $\hat{z}_i = z_i + r R_i e_2$, and that $\gamma_i = \beta_i - r$, we have

$$\hat{z}_i = z_i + r R_i e_2 = x_i + (\beta_i - r) R_i e_2 + r R_i e_2 = x_i + \beta_i R_i e_2 = \hat{x}_i, \tag{8.34}$$

[2] We have picked $\eta = \mathbf{1}$ in (4.15).

from which it follows that, in (x, θ) coordinates, the set $\hat{\Gamma}_{cf}$ is precisely Γ_{cf}. Thus, designing a feedback for the virtual unicycle team that stabilizes $\hat{\Gamma}_{cf}$ is equivalent to designing a feedback that stabilizes Γ_{cf} for (8.1). In turn, in (z, θ) coordinates, stabilizing $\hat{\Gamma}_{cf}$ corresponds to solving a version of CFP for the virtual unicycle team in which the formation is on a common circle of radius r (because the distance between $\hat{z}_i$ and z_i is r for all $i \in \mathbf{n}$), and the offset heading angles are $\rho_i(d, \beta_1)$. This is a problem we know how to solve by means of the feedback (8.31) (with the almost global enhancement mentioned earlier). Based on (8.31), we thus propose the feedback for the virtual unicycle team,

$$\hat{u}_i = \bar{u}_i(\varphi_i) = r\bar{v} + kg_i((\theta_{ij} - \rho_{ij}(d, \beta_1))_{j \in \mathcal{N}_i}, \mathbf{1}),$$
$$\omega_i = \omega_i^\star(\varphi_i, \nu_i(y_i^i)) = \frac{\bar{u}_i(\varphi_i)}{r} + K\hat{\varphi}(\nu_i(\hat{y}_i^i))\nu_i(\hat{y}_i^i), \; i \in \mathbf{n}.$$

In writing the above, we have used the fact that the function ν_i for system (8.32) is a function of the relative quantities $(\hat{z}_{ij}^i)_{j \in \mathcal{N}_i}$, but we have shown in (8.34) that $\hat{z}_i = \hat{x}_i$, which implies that the vector $(\hat{z}_{ij}^i)_{j \in \mathcal{N}_i}$ coincides with $\hat{y}_i^i$.

Using the fact that $u_i = \hat{u}_i + \gamma_i \omega_i = \hat{u}_i + (\beta_i - r)\omega_i$, we return to (x, θ) coordinates and get the feedback

$$\begin{aligned} u_i &= u_i^\star(\varphi_i, \nu_i(y_i^i)) = \bar{u}_i(\varphi_i) + (\beta_i - r)\omega_i^\star(\varphi_i, \nu_i(y_i^i)), \\ \omega_i &= \omega_i^\star(\varphi_i, \nu_i(y_i^i)) = \frac{\bar{u}_i(\varphi_i)}{r} + K\hat{\varphi}(\nu_i(\hat{y}_i^i))\nu_i(\hat{y}_i^i), \; i \in \mathbf{n}, \end{aligned} \tag{8.35}$$

where

$$\bar{u}_i(\varphi_i) = r\bar{v} + kg_i((\theta_{ij} - \rho_{ij}(d, \beta_1))_{j \in \mathcal{N}_i}, \mathbf{1}).$$

We will show in Theorem 8.2 that the feedback (8.35) does indeed solve CFP for the original unicycle team in (8.1).

The idea of using the virtual unicycle team to convert CFP to a problem that can be solved using the controller of [1] is depicted in Fig. 8.1.

Control design for CPP. Having designed a solution to CFP, we now need to adapt the feedback in (8.35) so as to stabilize the subset of Γ_{cf} in which $\hat{x}_1 = c$, where $c \in \mathbb{R}^2$ is the desired center of the concentric circles that the unicycles are to follow in steady state. For this, we leverage the fact that the vector $(c - x_i)^i$ is now assumed to be available for feedback, and replace the consensus controller in $\nu_i(y_i^i)$ in (8.35) by an equilibrium stabilizer. Specifically, we now define $\nu_i((c - x_i)^i) := -(c - \hat{x}_i)^i \cdot e_1$, and we will show in Theorem 8.2 that feedback (8.35) with this new definition of ν_i solves CPP.

A unified controller structure. To summarize, we have identified once again a unified controller structure that is capable of solving both CFP and CPP. We repeat it here for convenience

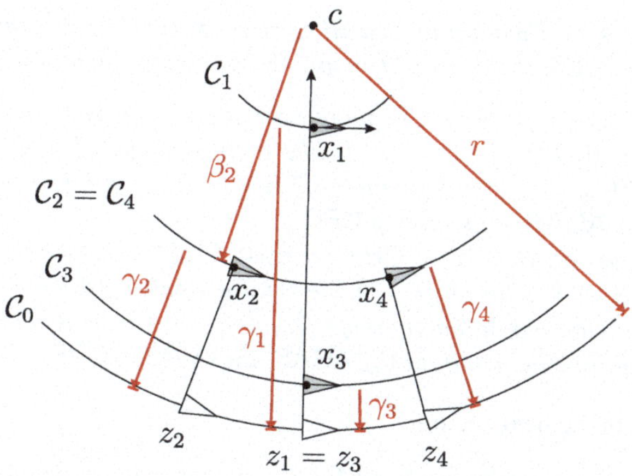

Fig. 8.1 Desired formation of unicycles (x, θ) (shaded) is achieved when (z, θ) (not shaded) lie on a common circle of radius r with the desired spacing $\theta_{1i} = \rho_i(d, \beta_1)$ for all $i \in \mathbf{n}$

Table 8.2 Auxiliary functions ν_i used to solve the formation problems with circular linear motion

	Problem	Choice of function ν_i
A	CFP	$\nu_i(\hat{y}_i^i) = -f(\hat{y}_i^i) \cdot e_1$, and $f(\cdot)$ is given in (4.2)
B	CPP	$\nu_i((c - x_i)^i) = -(c - \hat{x}_i)^i \cdot e_1$

$$
\begin{aligned}
u_i &= u_i^\star(\varphi_i, \nu_i) = \bar{u}_i(\varphi_i) + (\beta_i - r)\omega_i^\star(\varphi_i, \nu_i), \\
\omega_i &= \omega_i^\star(\varphi_i, \nu_i) = \frac{\bar{u}_i(\varphi_i)}{r} + K\hat{\varphi}(\nu_i)\nu_i, \ i \in \mathbf{n},
\end{aligned}
\tag{8.36}
$$

where

$$
\bar{u}_i(\varphi_i) = r\bar{v} + kg_i((\theta_{ij} - \rho_{ij}(d, \beta_1))_{j \in \mathcal{N}_i}, \mathbf{1}),
$$

and:

- $k, K > 0$ are design parameters;
- $g_i(\cdot, \mathbf{1})$ is the almost global rotational integrator consensus controller in (4.15) with $\eta = \mathbf{1}$;
- $\hat{\varphi}(\nu_i) := 1/(1 + \|\nu_i\|)$;
- ν_i is an auxiliary controller defined in Table 8.2.

The next result formalizes the discussion above and states that the controller structure in (8.36) with auxiliary functions ν_i in Table 8.2 does indeed solve the two formation control problems with final circular motion.

Theorem 8.2 (Solutions to CFP *and* CPP) *Consider the team of unicycles in* (8.1) *with connected, undirected sensor graph $\mathcal{G}$ and any parameters $a_{ij} = a_{ji} > 0$ for*

$i \in \mathbf{n}$, $j \in \mathcal{N}_i$ *in* (4.2). *Then, for almost all parameters (in the sense of Theorem 4.2)* $b_{ij} = b_{ji} > 0$ *in* (4.15), *the control structure* (8.36) *enjoys the following properties.*

(i) *(*CFP*) For any* $(d, \beta_1, \bar{v}) \in$ CF, *there exist* $K^\star, k^\star > 0$ *such that for any* $K \in (0, K^\star)$, $k \in (0, k^\star)$, *for any* $r > 0$, *the local and distributed control inputs in* (8.36) *with* v_i *as in row A of Table* 8.2 *render the circular formation flocking manifold* Γ_{cf} AGAS, *thus solving* CFP.
(ii) *(*CPP*) For any* $(d, \beta_1, c, \bar{v}) \in$ CP *there exist* $K^\star, k^\star > 0$ *such that for any* $K \in (0, K^\star)$, $k \in (0, k^\star)$, *for any* $r > 0$, *the control inputs in* (8.36) *with* v_i *as in row B of Table* 8.2 *are strictly functions of* $(y_i^i, \varphi_i, (c - x_i)^i)$ *and render the formation circle path following manifold* Γ_{cp} AGAS, *thus solving* CPP.

The proof is in the next section.

Remark 8.3 The control inputs in (8.36) for unicycle i are strictly functions of (y_i^i, φ_i), the formation parameters $(\beta_j)_{j\in\mathcal{N}_i}$ and v_i. In the case of flocking, v_i is a function of y_i^i, while in the case of path following, v_i is a function of $(c - x_i)^i$ because $(c - \hat{x}_i)^i = (c - x_i)^i - \beta_i e_2$. Moreover, it is readily seen that $(u_i^\star, \omega_i^\star)|_{\Gamma_{cf}} = (u_i^\star, \omega_i^\star)|_{\Gamma_{cp}} = (\bar{v}\beta_i, \bar{v})$, as required by CFP and CPP.

Remark 8.4 The solutions presented in Theorems 8.1 and 8.2 for line and circle path following assume that all unicycles in the ensemble sense the desired path. A more general result can be obtained for the case where only a nonempty subset of unicycles $S \subset \mathcal{V}$ sees the path assuming that at least one unicycle in S is a globally reachable node for $\mathcal{G}$. For undirected graphs, S can be any nonempty subset of $\mathcal{V}$. For line path following, choose the feedback law as in (8.10) with $\mu_i^i = \bar{v}p^i$ for $i \notin S$ and $\mu_i^i = h(\hat{x}_i)^i = k_0(c^\star(\hat{x}_i) - \hat{x}_i)^i + \bar{v}p^i$ for $i \in S$. Measurement of p^i is still required by all agents in this case. For circle path following choose the feedback law as in (8.36) with $v_i = -f(\hat{y}_i^i) \cdot e_1$ for $i \notin S$, and $v_i = -f(\hat{y}_i^i) \cdot e_1 - (c - \hat{x}_i)^i \cdot e_1$ for $i \in S$.

8.5 Proof of Theorem 8.2

Recall the definition of z_i and $\hat{z}_i$ from the previous section: $z_i = x_i + \gamma_i R_i e_2$, with $\gamma_i = \beta_i - r$, and $\hat{z}_i = z_i + r R_i e_2 = \hat{x}_i$. The last identity was shown in (8.34). Denote $z = (z_i)_{i\in\mathbf{n}}$ and $\hat{z} = (\hat{z}_i)_{i\in\mathbf{n}}$. Further, define the relative coordinates $(\tilde{z}, \tilde{\theta})$ where $\tilde{z} := (\hat{z}_{1i})_{i\in\mathbf{2:n}} \in \mathbb{R}^{2(n-1)}$ and $\tilde{\theta} := (\theta_{1i})_{i\in\mathbf{2:n}} \in \mathbb{T}^{n-1}$.

In (8.32), we have shown that

$$\begin{aligned} \dot{\hat{z}}_i &= \hat{u}_i R_i e_1, \\ \dot{\theta}_i &= \omega_i, \quad i \in \mathbf{n}, \end{aligned} \tag{8.37}$$

where $\hat{u}_i = u_i - \gamma_i \omega_i$. As we have seen in the previous section, the feedback (8.36) turns into the following feedback for the system above

$$\begin{aligned} \hat{u}_i &= \bar{u}_i(\varphi_i) = r\bar{v} + kg_i((\theta_{ij} - \rho_{ij}(d, \beta_1))_{j\in\mathcal{N}_i}, \mathbf{1}), \\ \omega_i &= \omega_i^\star(\varphi_i, \nu_i(y_i^i)) = \frac{\bar{u}_i(\varphi_i)}{r} + K\hat{\varphi}(\nu_i)\nu_i, \; i \in \mathbf{n}. \end{aligned} \tag{8.38}$$

Since $\hat{u}_i$ and ω_i are bounded, the equations on the right-hand side of (8.37) are bounded, and therefore, the closed-loop system has no finite escape times. Substituting (8.38) into (8.37) and expressing the resulting dynamics in $(\hat{z}, \theta)$ coordinates, one gets

$$\begin{aligned} \dot{\hat{z}}_i &= -rK\hat{\varphi}(\nu_i)\nu_i R_i e_1, \\ \dot{\theta}_i &= \frac{\bar{u}_i}{r} + K\hat{\varphi}(\nu_i)\nu_i, \; i \in \mathbf{n}. \end{aligned} \tag{8.39}$$

Since $\hat{x}_i = \hat{z}_i$, in $(\hat{z}, \theta)$ coordinates the target sets Γ_{cf} and Γ_{cp} read, respectively, as

$$\begin{aligned} \hat{\Gamma}_{cf} &:= \{(\hat{z}, \theta) \in \mathbb{R}^{2n} \times \mathbb{T}^n : \hat{z}_i = \hat{z}_1, \; \theta_{1i} = \rho_i(d, \beta_1), \; i \in \mathbf{2}{:}\mathbf{n}\}, \\ \hat{\Gamma}_{cp} &:= \{(\hat{z}, \theta) \in \hat{\Gamma}_{cf} : \hat{z}_1 = c\}. \end{aligned}$$

Proof of CFP

We begin by proving that $\hat{\Gamma}_{cf}$ is AGAS for (8.39), and for this, we adapt the proof technique from [1]. Let

$$\bar{\Gamma} := \{(\hat{z}, \theta) \in \mathbb{R}^{2n} \times \mathbb{T}^n : \hat{z}_i = \hat{z}_1\}, \tag{8.40}$$

and note that $\hat{\Gamma}_{cf} \subset \bar{\Gamma}$. Let ν_i be as in row A of Table 8.2 and make use of the fact that $\hat{y}_i^i = (\hat{x}_{ij}^i)_{j\in\mathcal{N}_i} = (\hat{z}_{ij}^i)_{j\in\mathcal{N}_i}$:

$$\nu_i((\hat{z}_{ij}^i)_{j\in\mathcal{N}_i}) = -f((\hat{z}_{ij}^i)_{j\in\mathcal{N}_i}) \cdot e_1.$$

In [1, Proposition V.2], it was proved that letting $K^\star = \bar{v}/2$, for each $K \in (0, K^\star)$, the feedback $\nu_i((\hat{z}_{ij}^i)_{j\in\mathcal{N}_i})$ above renders the set $\bar{\Gamma}$ GAS for (8.39).

We now consider the dynamics in $(\tilde{z}, \tilde{\theta})$ coordinates. We have just established that, for $K \in (0, K^\star)$, the set $\{(\tilde{z}, \tilde{\theta}) : \tilde{z} = 0\}$ is GAS. The dynamics of (8.39) on this set are described in $(\tilde{z}, \tilde{\theta})$ coordinates by

$$\begin{aligned} &\dot{\hat{z}}_{1i} = 0, \\ &\dot{\theta}_{1i} = (k/r)g_i((\theta_{ij} - \rho_{ij}(d, \beta_1))_{j\in\mathcal{N}_i}, \mathbf{1}) - (k/r)g_1((\theta_{1j} - \rho_j(d, \beta_1))_{j\in\mathcal{N}_1}, \mathbf{1}). \end{aligned} \tag{8.41}$$

By Theorem 4.2, the rotational part of system (8.41) has a globally attractive compact set of isolated equilibria,

$$\mathcal{A} \cup \left\{\tilde{\theta} \in \mathbb{T}^{n-1} : \theta_{1j} = \rho_j(d, \beta_1), \; j \in \mathbf{2}{:}\mathbf{n}\right\} = \mathcal{A} \cup \{\rho(d, \beta_1)\},$$

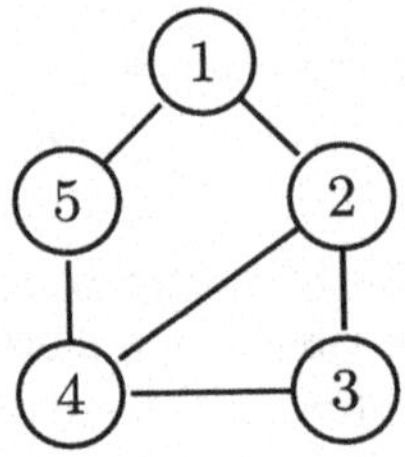

Fig. 8.2 Undirected graph $\mathcal{G}$ under consideration in the simulation results

in which the equilibrium $\{\rho(d, \beta_1)\}$ is asymptotically stable while the equlibria in $\mathcal{A}$ are exponentially unstable. By the novel reduction theorem for almost global asymptotic stability, Theorem A.6, in $(\tilde{z}, \tilde{\theta})$ coordinates the set $\{(\tilde{z}, \tilde{\theta}) : \tilde{z} = 0, \tilde{\theta} = \rho(d, \beta_1)\}$ is AGAS. Going back to $(\hat{z}, \theta)$ coordinates, since system (8.39) has no finite escape times, the set $\hat{\Gamma}_{cf}$ is AGAS for system (8.39). This in turn implies that the set Γ_{cf} is AGAS for system (8.1) with the feedback in (8.36). This proves part (i) of the theorem.

Proof of CPP

We now prove that $\hat{\Gamma}_{cp}$ is AGAS for (8.39). Consider again system (8.39), but this time setting $v_i = -(c - \hat{x}_i) \cdot R_i e_1$. Define a new set $\bar{\Gamma}$ (this takes the place of the set $\bar{\Gamma}$ defined earlier) as

$$\bar{\Gamma} := \{(\hat{z}, \theta) \in \mathbb{R}^{2n} \times \mathbb{T}^n : \hat{z}_i = c, \ i \in \mathbf{n}\},$$

so that $\hat{\Gamma}_{cp} \subset \bar{\Gamma}$. We claim that $\bar{\Gamma}$ is GAS. Define the Lyapunov function $V = \sum_{i=1}^{n} \|c - \hat{x}_i\|^2$ with derivative given by

$$\begin{aligned} \dot{V} &= \sum_{i=1}^{n} r K \hat{\varphi}(v_i) v_i (c - \hat{x}_i) \cdot R_i e_1 \\ &= \sum_{i=1}^{n} -r K \hat{\varphi}(v_i) \|v_i\|^2 \leq 0. \end{aligned}$$

Since $\dot{V} \leq 0$, and since $V^{-1}(0) = \bar{\Gamma}$, the set $\bar{\Gamma}$ is stable. To prove global attractivity, we begin by observing that the level sets of V are compact in $(\hat{x}, \theta)$ coordinates, and the Krasovskii–LaSalle invariance principle implies that $v_i \to 0$ as $t \to \infty$. Since

$$\dot{\theta}_i = \frac{\bar{u}_i}{r} + K \hat{\varphi}(v_i) v_i,$$

and since $\|\hat{\varphi}(v_i) v_i\| < 1$, we have

$$\dot{\theta}_i \geq \frac{\inf_{\theta \in \mathbb{T}^n} \bar{u}_i(\theta)}{r} - K.$$

Referring to the definition of $\bar{u}_i$ in (8.38), the function g_i is uniformly bounded, being continuous and defined on a compact set. This implies that there exists $k^\star > 0$ such that for each $k \in (0, k^\star)$, $\inf_{\theta \in \mathbb{T}^n} \bar{u}_i(\varphi_i) > r\bar{v}/2$. Moreover, $\sup_{\theta \in \mathbb{T}^n} \bar{u}_i < \infty$ for all $i \in \mathbf{n}$. Then, there exist $\zeta_1, \zeta_2 > 0$ such that $0 < \zeta_1 < \dot{\theta}_i < \zeta_2$ for all $i \in \mathbf{n}$, for all time and initial conditions. Using this property, and using ideas from the proof of [1, Proposition V.2], one can show that $(\hat{x}_i - c) \to 0$ as $t \to \infty$. Together with stability, this implies that $\hat{\Gamma}_2$ is GAS, as claimed. We also deduce boundedness of the states $(\hat{x}, \theta)$. On the set $\bar{\Gamma}$, we have $v_i = 0$ for all $i \in \mathbf{n}$, and the dynamics of θ_i reduce to

$$\dot{\theta}_i = \bar{v} + (k/r)g_i((\theta_{ij} - \rho_{ij}(d, \beta_1))_{j \in \mathcal{N}_i}, \mathbf{1}).$$

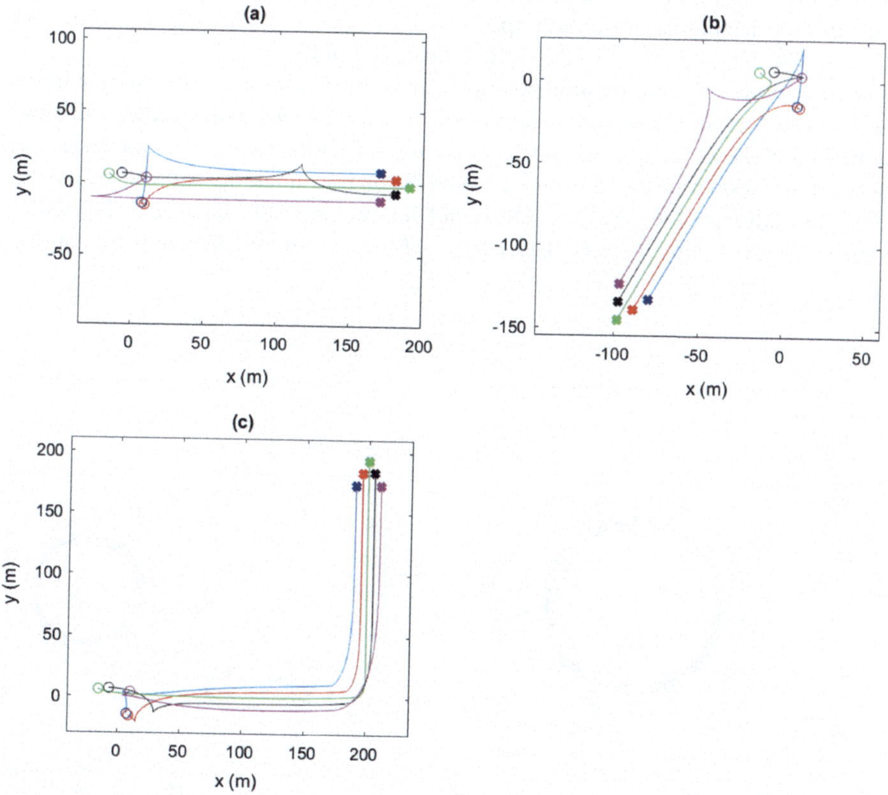

Fig. 8.3 Simulation results for formations with final parallel collective motion: **a** formation flocking with beacon $p = (1, 0)$ pointing in the direction of the positive x-axis, **b** formation flocking with no beacon where $\mu_i^i = (\bar{v}/|\mathcal{N}_i|) \sum_{j \in \mathcal{N}_i} R_j^i e_1,\ i \in \mathbf{n}$, **c** formation path following for $\mathcal{C}(r_0, p) = \{x \in \mathbb{R}^2 : x = r_0 + sp,\ s \in \mathbb{R}\}$ with $r_0 = (200, 0)$, $p = (0, 1)$. Initial positions are indicated with $\circ$ and positions at the end of the simulation are indicated with $\times$

Invoking again Theorems 4.2 and A.6, and repeating the arguments we made above, we conclude that the set $\hat{\Gamma}_{cp}$ is AGAS. This in turn implies that the set Γ_{cp} is AGAS for system (8.1) with the feedback in (8.36). This concludes the proof of the theorem. □

8.6 Simulation Results

In this section, we present simulation results for formation control with parallel and circular final collective motions using the feedbacks presented in Sects. 8.2 and 8.4. We consider a group of five unicycles with undirected graph shown in Fig. 8.2 and with desired triangular formation specified by $d^1_{12} = (-10, 5)$, $d^1_{13} = (-10, -5)$, $d^1_{12} = (-20, 10)$ and $d^1_{12} = (-20, -10)$, illustrated in Fig. 7.6.

We have chosen random initial positions on a 40m × 40m area with random initial angles. The corresponding simulation results for formations with parallel collective motions are shown in Fig. 8.3 while those for formations with circular collective motions are shown in Fig. 8.4. For parallel formation flocking with no beacon in Fig. 8.3b, although the sensor graph is not hierarchical, the local and distributed control in (8.10) with $\mu^i_i = (\bar{v}/|\mathcal{N}_i|) \sum_{j\in\mathcal{N}_i} R^i_j e_1$, $i \in \mathbf{n}$ still manages to stabilize Γ_{pf}.

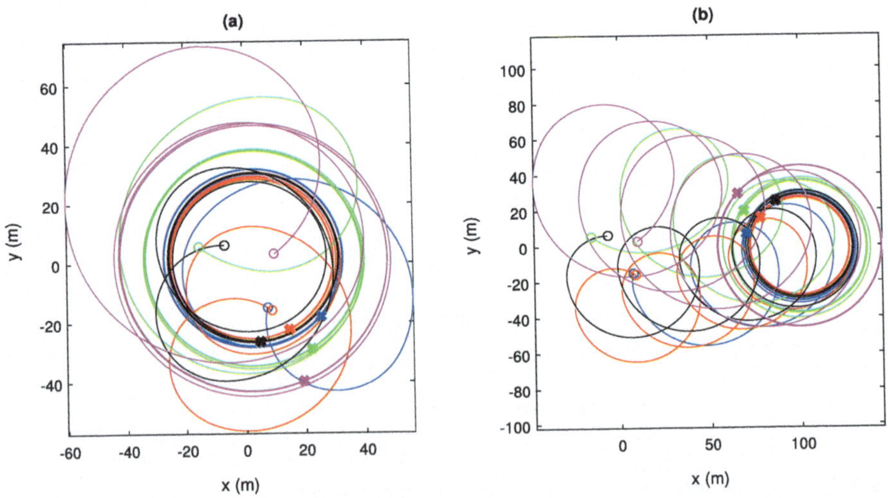

Fig. 8.4 Simulation for formations with circular collective motion **a** formation flocking, **b** formation path following around the point $c = (100, 0)$. Initial positions are indicated with ○, and positions at the end of the simulation are indicated with ×

Reference

1. El-Hawwary, M., Maggiore, M.: Distributed circular formation stabilization for dynamic unicycles. IEEE Trans. Autom. Control **58**(1), 149–162 (2013)

Chapter 9
Unicycle Formation Simulation Trials

Abstract In this chapter, we present a set of simulation trials with the intention of testing the feedback solving the parallel formation problem PP in (7.4). We investigate a number of scenarios which do not satisfy one or more assumptions in Theorem 7.1:

- The sensor graph is state dependent, as opposed to static, which corresponds to the situation when each unicycle has a limited sensing radius.
- The sensor graph is directed as opposed to undirected.
- The high-gain conditions on $\bar{\alpha}$ and k are ignored and low gains are used.
- Unmodeled factors including sensor noise, input noise, sampling and saturated inputs affecting the unicycle team are included.

We will not present additional simulations for formations with linear and circular final collective motions discussed in Chap. 8. However, we predict similar outcomes since the solutions in this chapter are based on a similar methodology.

9.1 Performance Measures

In this section, we introduce a number of performance measures that will be used to evaluate the parallel formation feedback (7.4). These include a formation measure, a drift measure, a collision measure and input measures.

Formation Measure. The most important measure of performance for the formation control problem is whether or not the unicycles achieve the desired formation. Any test trial that does not meet this condition is considered a failure. This aspect can be captured by a suitable *formation measure*. We define a formation measure $\mu_f : \mathbb{R}^{2n} \times \mathbb{T}^n \to \mathbb{R}$ satisfying two conditions:

1. μ_f is positive semidefinite, i.e., $\mu_f \geq 0$;
2. μ_f equals zero if and only if the formation is achieved, i.e., $\mu_f^{-1}(0) = \Gamma_p$,

where Γ_p is the parallel formation manifold defined in (3.4).

A. Roza et al., *Distributed Coordination Theory for Robot Teams*,
Lecture Notes in Control and Information Sciences 490,
https://doi.org/10.1007/978-3-030-96087-2_9

We have seen in Chap. 7 that achieving formation amounts to the simultaneous synchronization of the heading angles $(\theta_i)_{i\in\mathbf{n}}$ and the endpoints $(\hat{x}_i)_{i\in\mathbf{n}}$, where $\hat{x}_i = x_i + \alpha_i R_i e_1 + \beta_i R_i e_2$, so that the parallel formation manifold can be expressed as

$$\Gamma_p = \left\{(x,\theta) \in \mathbb{R}^{2n} \times \mathbb{T}^n : \theta_i = \theta_1, \hat{x}_i = \hat{x}_1, i \in \mathbf{2}:\mathbf{n}\right\}.$$

Consider a collection of n unicycles, and define an average endpoint $\bar{\hat{x}} \in \mathbb{R}^2$ and average angle $\bar{\theta} \in \mathbb{S}^1$ as

$$\bar{\hat{x}} := \frac{1}{n}\sum_{i=1}^{n} \hat{x}_i, \quad \bar{\theta} := \operatorname{atan2}\left(\sum_{i=1}^{n} \sin\theta_j, \sum_{i=1}^{n} \cos\theta_j\right).$$

The above average angle $\bar{\theta}$ has a singularity when $\left(\sum_{i=1}^{n} \sin\theta_j, \sum_{i=1}^{n} \cos\theta_j\right) = (0,0)$. The parallel formation manifold may be expressed in terms of $(\bar{x}, \bar{\theta})$ as

$$\Gamma_p = \left\{(x,\theta) \in \mathbb{R}^{2n} \times \mathbb{T}^n : \hat{x}_i = \bar{\hat{x}},\ \theta_i = \bar{\theta}_i,\ i \in \mathbf{n}\right\}.$$

See Fig. 9.1 for an illustration of $(\bar{\hat{x}}, \bar{\theta})$. The desired position of unicycle i, which we denote by $x_{i,des}$, can be expressed in terms of the average position $\hat{x}$ as,

$$x_{i,des} = \bar{\hat{x}} - \alpha_i \bar{R} e_1 - \beta_i \bar{R} e_2,$$

where $\bar{R}$ is the rotation matrix in $\mathsf{SO}(2)$ with angle $\bar{\theta}$. This is illustrated in Fig. 9.1.

The error between x_i and $x_{i,des}$ is

$$x_i - x_{i,des} = (\hat{x}_i - \bar{\hat{x}}) + (\alpha_i R_i e_1 + \beta_i R_i e_2 - \alpha_i \bar{R} e_1 + \beta_i \bar{R} e_2),$$

and referring to Fig. 9.1, the norm of the vector $(\alpha_i R_i e_1 + \beta_i R_i e_2 - \alpha_i \bar{R} e_1 + \beta_i \bar{R} e_2)$ is

$$\|\alpha_i R_i e_1 + \beta_i R_i e_2 - \alpha_i \bar{R} e_1 + \beta_i \bar{R} e_2\| = 2\sqrt{\alpha_i^2 + \beta_i^2}\left\|\sin\left(\frac{\theta_i - \bar{\theta}}{2}\right)\right\|,$$

from which we deduce that

$$\|x_i - x_{i,des}\| < \|\hat{x}_i - \bar{\hat{x}}\| + 2\sqrt{\alpha_i^2 + \beta_i^2}\left\|\sin\left(\frac{\theta_i - \bar{\theta}}{2}\right)\right\| =: e_i(x,\theta).$$

Letting $e_{\max} := \max_{i\in\mathbf{n}} e_i$ and $d_{\max} := \max_{i\in\mathbf{2}:\mathbf{n}} \|d_{1i}^1\|$, we define the formation measure as

$$\mu_f(x,\theta) := \frac{e_{\max}}{d_{\max}}.$$

It is not hard to see that $\mu_f(x,\theta) = 0$ if and only if $(x,\theta) \in \Gamma_p$, and so μ_f satisfies the second property of the formation measure.

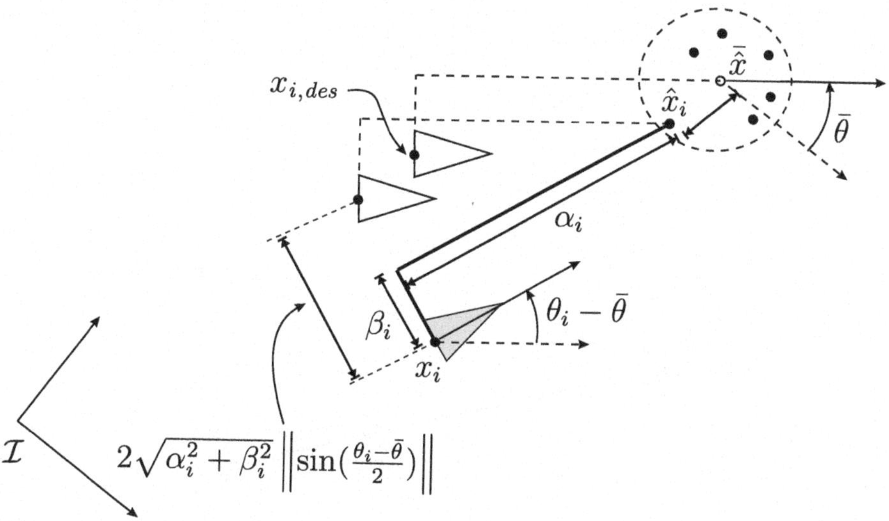

Fig. 9.1 Illustration of the formation measure. The endpoints $(\hat{x}_i)_{i \in \mathbf{n}}$ are indicated by black dots, and their average $\bar{\hat{x}}$ is indicated by ∘. The angle $\bar{\theta}$ represents the average heading angle of the unicycles with respect to the inertial frame and is indicated by the solid arrow with tail at $\bar{\hat{x}}$. The point $x_{i,des}$ represents the position that unicycle i should have to be in a formation whose average offset position and heading angle are given by $(\bar{\hat{x}}, \bar{\theta})$

Note that $e_{\max}$ represents a bound on the formation error, while $d_{\max}$ quantifies the *size of the formation*. The measure μ_f just defined is thus normalized by the size of the formation, which allows us to use the same measure to compare the performance of the control strategy for arbitrary formations. We say that a formation is achieved if there exists a time t_f after which the formation measure satisfies $\mu_f < \bar{\mu}_f$ where $\bar{\mu}_f > 0$ is called the *formation threshold*. Otherwise, the test is considered a failure.

Drift Measure. In an ideal scenario, a formation should be achieved without translating or drifting too much from its initial configuration. For example, if the unicycles are constrained to move within an enclosed building, drifting too much might cause the unicycles to collide with the walls before the formation is achieved. We quantify the drift with a *drift measure*, which compares the average position of unicycles at the initial time $\bar{x}(t_0) = (1/n)\sum_{i \in \mathbf{n}} x(t_0)$ to the average position of unicycles at a final time $\bar{x}(t_f) = (1/n)\sum_{i \in \mathbf{n}} x(t_f)$ where t_f is the time when the formation is achieved (i.e., when the formation measure satisfies $\mu_f < \bar{\mu}_f$). The drift measure is therefore defined as

$$\mu_d := \|\bar{x}(t_f) - \bar{x}(t_0)\|,$$

and is illustrated in Fig. 9.2.

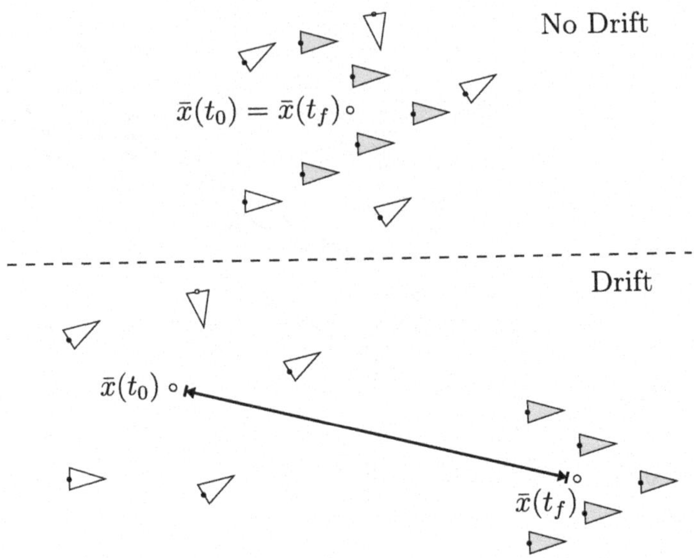

Fig. 9.2 Illustration of the drift measure. The unicycles at the initial time t_0 are not shaded while the unicycles at time t_f are shaded. The average position of the unicycles is indicated with $\circ$. In the top figure, $\bar{x}(t_0)$ and $\bar{x}(t_f)$ are the same and therefore $\mu_d = 0$. In the bottom figure, $\|\bar{x}(t_f) - \bar{x}(t_0)\| \neq 0$, and therefore, drift is present which is roughly twice the size of the formation itself

Collision Measure. The *collision measure* μ_c is the minimum distance between any two unicycles attained between the initial time t_0 and the final time t_f. Formally, it is defined as

$$\mu_c := \min_{t \in [t_0, t_f], i, j \in \mathbf{n}} \|x_{ij}(t)\|.$$

Assuming robots are 0.5 m in radius or less, we say that a collision occurs if $\mu_c < 0.5$.

Input Measures. Finally, we define four *input measures* that correspond to the supremum and the mean of the magnitudes of the control inputs, achieved in the time interval $[t_0, t_f]$ and maximized over $i \in \mathbf{n}$. The input measures are:

- $\mu_{u,1} := \max_{i \in \mathbf{n}} \sup_{t \in [t_0, t_f]} |u_i(t)|$ (m/s);
- $\mu_{\omega,1} := \max_{i \in \mathbf{n}} \sup_{t \in [t_0, t_f]} |\omega_i(t)|$ (rad/s);
- $\mu_{u,2} := \max_{i \in \mathbf{n}} \text{mean}_{t \in [t_0, t_f]}(|u_i(t)|)$ (m/s);
- $\mu_{\omega,2} := \max_{i \in \mathbf{n}} \text{mean}_{t \in [t_0, t_f]}(|\omega_i(t)|)$ (rad/s).

9.2 Simulation Trials

In this section, we present several simulation trials to test the control solution (7.4) under different parameter choices. Each trial is composed of $N = 500$ simulations. As the target formation, we choose the same five-unicycle formation used in the sim-

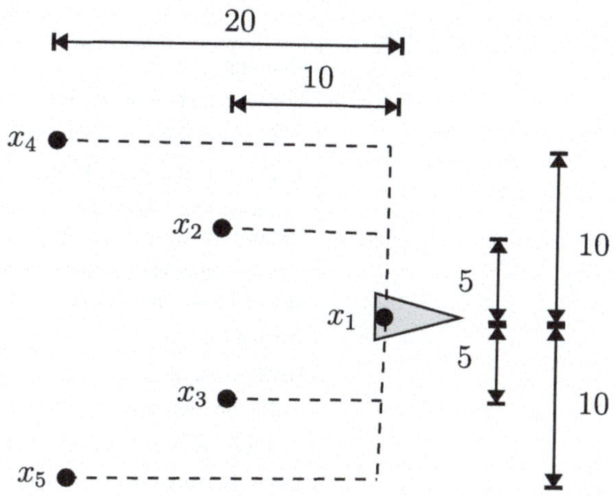

Fig. 9.3 Triangular formation specified by the offset vectors $d^1_{12} = (-10, 5)$, $d^1_{13} = (-10, -5)$, $d^1_{14} = (-20, 10)$ and $d^1_{15} = (-20, -10)$

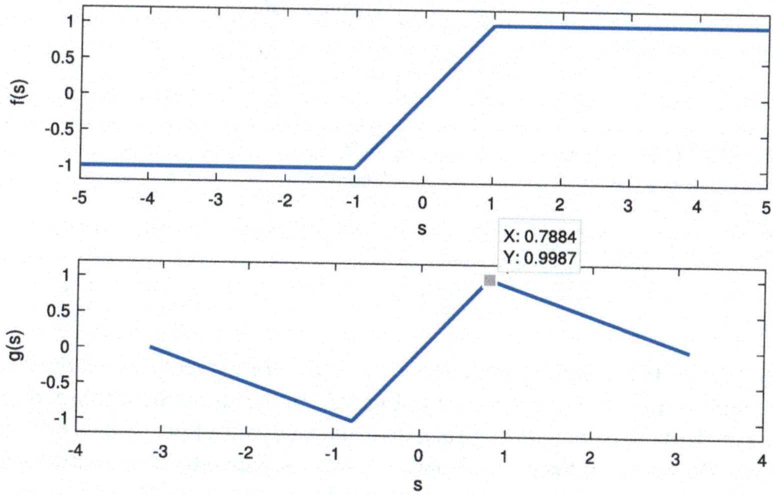

Fig. 9.4 Interaction function $f(s)$ and $g(s)$

ulation trials of Chap. 7. The formation is illustrated in Fig. 9.3. The initial angles $(\theta_i)_{i\in\mathbf{n}}$ are chosen randomly in $\mathbb{T}^n$, and the positions $(x_i)_{i\in\mathbf{n}}$ are chosen randomly in a 60×60 meter grid. We choose the interaction function $f(s)$ in (4.3) and rotational interaction function $g(s)$ in (4.15) as shown in Fig. 9.4. We set $a_{ij} = 30$ and $b_{ij} = \alpha_i + \alpha_j$ for all $j \in \mathcal{N}_i$, $\eta_i = 1/\alpha_i$, and $t_0 = 0$. For each test, we compute the quantities in Table 9.1.

We begin by presenting a set of simulations that will serve as a reference for the remaining test trials. From here on, we will refer to this set of simulations as the

Table 9.1 Data collected from simulations

Quantity	Meaning
C	# of tests that begin with a connected graph
D	# of tests that begin with a disconnected graph
F	# of tests that achieve formation
CF	# of tests that begin with a connected graph and end in formation ($\mu_f < \bar{\mu}_f$)
DF	# of tests that begin with a disconnected graph and end in formation ($\mu_f < \bar{\mu}_f$)
t_f	Formation time
μ_d	Drift measure
μ_c	Collision measure
Nc	# of tests in which a collision occurs
$\mu_{u,1}, \mu_{\omega,1}, \mu_{u,2}, \mu_{\omega,2}$	Input measures

Table 9.2 Reference case parameters

Quantity	Reference value
$\bar{\mu}_f$ (formation threshold)	0.05
k	15
$\bar{\alpha}$	5
Fixed or state-dependent graphs	Fixed
Undirected or directed graph	Undirected
Input saturation	No
Sampling and disturbances	No

reference case. The control parameters $(\bar{\mu}_f, k, \bar{\alpha})$ are presented in Table 9.2. The sensor graph is undirected and shown in Fig. 9.6. In the subsequent subsections, we will vary each of the parameters one at a time and study the effect on the performance measures. Moreover, in Sects. 9.2.3 and 9.2.4 we will also test the control solution under different assumptions on the sensor graph.

The results of the simulations are summarized in Table 9.3. The column called F/N (%) represents the percentage of simulations in which the formation is achieved, i.e., when there exists a time t_f after which $\mu_f < 0.05$. The column called Nc/N (%) represents the percentage of simulations in which a collision occurs. The remaining columns correspond to the performance measures. The quantity $t_f \mu_{u,2}$ is an upper bound of the distance traveled by any single robot in the ensemble and $t_f \mu_{\omega,2}$ is an upper bound of the amount of rotation that any robot performs before formation is achieved. The quantity $t_f \mu_{\omega,2}/2\pi$ is an upper bound of the number of revolutions spun by any robot. The values presented in Table 9.3 are averages taken over the $N = 500$ iterations.

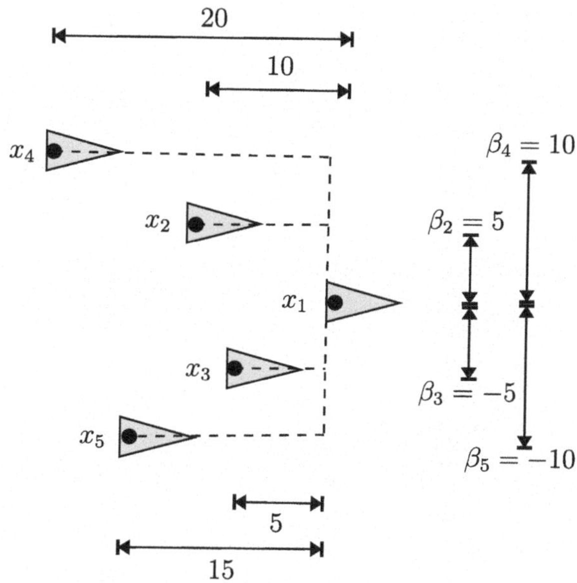

Fig. 9.5 Formation specified by $d_{12}^1 = (-10, 5)$, $d_{13}^1 = (-5, -5)$, $d_{14}^1 = (-20, 10)$ and $d_{15}^1 = (-15, -10)$

Table 9.3 Reference case simulation results

F/N (%)	μ_d (m)	Nc/N (%)	t_f (s)		
100	76.4330	68.20	220.3635		
$\mu_{u,1}$ (m/s)	$\mu_{\omega,1}$ (rad/s)	$\mu_{u,2}$ (m/s)	$\mu_{\omega,2}$ (rad/s)	$t_f \mu_{u,2}$ (m)	$t_f \mu_{\omega,2}$ (rad)
76.4238	17.1043	0.5888	0.0187	129.7534	4.1278

From this first set of simulations, we can draw the following conclusions:

- F/N (%)= 100 indicates that all simulations achieve formation.
- The average drift of the formation is 76 m, i.e., 3.4 times the formation size.
- There is a collision 68% of the time which is high. To avoid this, one would have to design a high-level collision avoidance layer. It turns out that the colliding agents are precisely those sharing common α_i values. If one considers, instead, the formation in Fig. 9.5, where no two unicycles share a common α_i value, there is a collision only 11% of the time.
- The average time to achieve formation is 220 s.
- The maximum speed input is 76.4238 m/s, and the maximum angular speed input is 17.1043 rad/s. These are both large but can be resolved by using input saturation as we discuss later.
- The maximum displacement of any single unicycle in the ensemble is 129.7534 m, and the maximum number of revolutions of any single unicycle is $4.1278/2\pi = 0.66$. Therefore, the solution has very little oscillation, which is a desirable property (Fig. 9.6).

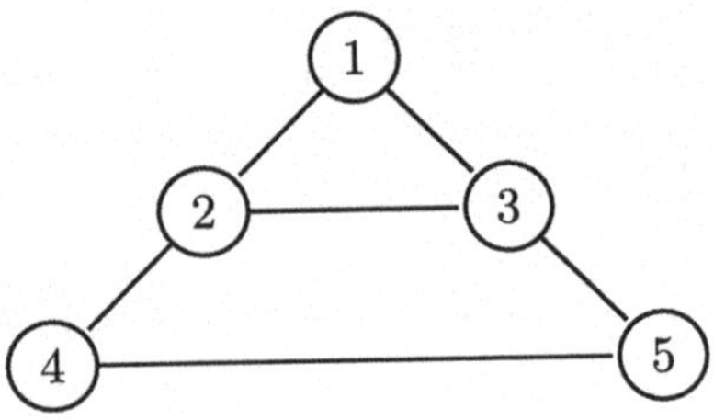

Fig. 9.6 Undirected graph used in the reference case

Table 9.4 Simulation results varying $\bar{\mu}_f$

$\bar{\mu}_f$	μ_d (m)	t_f (s)
0.01	115.0104	5794.5
0.02	99.0958	1403.7
0.05	**76.4330**	**220.3635**
0.1	60.0024	54.0723
0.15	51.3388	24.1076
0.2	43.0169	13.1582
0.25	39.2520	8.9690

9.2.1 *Variation of the Formation Threshold*

In this section, we will show how changing the formation threshold $\bar{\mu}_f$ affects the performance measures μ_d and t_f. The average results for $N = 500$ simulations are shown in Table 9.4. The bold values correspond to the reference simulations. We notice that both μ_d and t_f are inversely proportional to $\bar{\mu}_f$. We can see that the smaller the threshold values of $\bar{\mu}_f$, the further the formation drifts. Also, the time to achieve smaller formation threshold values grows very rapidly. To speed up convergence in a neighborhood of the formation (where the rate of convergence is slow), one can always add a gain K to the controller in (7.4) as follows,

$$\begin{aligned} u_i^\star(y_i^i, \varphi_i) &= K f_i(\hat{y}_i^i) \cdot e_1 + \beta_i \omega_i^\star(y_i^i, \varphi_i), \\ \omega_i^\star(y_i^i, \varphi_i) &= \frac{K}{\alpha_i} \left(f_i(\hat{y}_i^i) \cdot e_2 + k g_i(\varphi_i, \eta) \right), \quad i \in \mathbf{n}. \end{aligned} \tag{9.1}$$

Adding a gain will increase the rate of convergence by a factor of K. However, this will also increase the magnitude of the control inputs by a factor of K and one may therefore need to saturate the input. In Fig. 9.7, we illustrate formations (relative to the frame of unicycle 1) obtained for different values of $\bar{\mu}_f$.

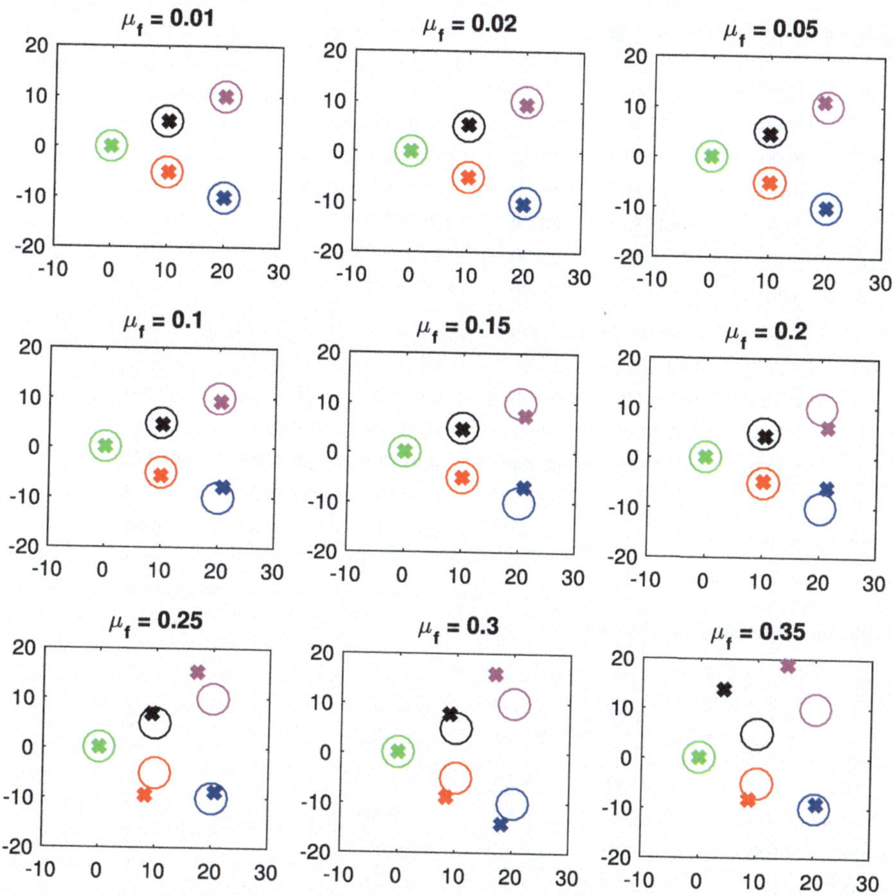

Fig. 9.7 Illustration of configurations satisfying different values of μ_f. The center of the circles represent the desired unicycle positions while the actual unicycle positions are indicated with $\times$

9.2.2 *Variation of the High-Gain Parameters $\bar{\alpha}$ and k*

Based on Theorem 7.1, when designing controller (7.4), we need to choose $\bar{\alpha}$ and k as high-gain parameters. In this section, we show that, in simulations, these gains do not actually need to be chosen too large to still achieve formation. The simulation results varying $\bar{\alpha}$ are presented in Table 9.5, and the results varying k are presented in Table 9.6. The values of the reference case are in bold.

The main conclusions from the simulation results in Table 9.5, obtained by varying $\bar{\alpha}$ are:

- For $\bar{\alpha} < 5$ (the reference value), some simulations fail. Only 2.2% of simulations fail when $\bar{\alpha}$ is chosen as low as 0.1.
- The drift is proportional to $\bar{\alpha}$.

Table 9.5 Simulation results varying $\bar{\alpha}$

$\bar{\alpha}$	F/N (%)	μ_d (m)	Nc/N (%)	t_f (s)		
0.1	97.80	65.8306	61.76	196.4190		
0.5	98.00	69.6150	66.33	192.7916		
1	99.40	68.6228	64.59	187.5087		
5	**100**	**76.4330**	**68.20**	**220.3635**		
10	100	89.1039	70.00	315.9337		
25	100	138.9444	69.20	853.3169		
$\bar{\alpha}$	$\mu_{u,1}$ (m/s)	$\mu_{\omega,1}$ (rad/s)	$\mu_{u,2}$ (m/s)	$\mu_{\omega,2}$ (rad/s)	$t_f\mu_{u,2}$ (m)	$t_f\mu_{\omega,2}$ (rad)
0.1	85.4778	994.4596	0.7337	0.0508	144.1049	9.9732
0.5	84.3687	660.0438	0.7010	0.0312	135.1392	6.0096
1	81.4894	197.9675	0.6934	0.0290	130.0245	5.4343
5	**76.4238**	**17.1043**	**0.5888**	**0.0187**	**129.7534**	**4.1278**
10	74.8656	7.6092	0.4822	0.0120	152.3422	3.7764
25	73.5076	3.2741	0.2501	0.0034	213.4184	2.9089

Table 9.6 Simulation results varying k

k	F/N (%)	μ_d (m)	Nc/N (%)	t_f (s)		
1	99.60	73.4022	67.07	2994.8		
5	99.80	73.8609	74.75	609.7329		
15	**100**	**76.4330**	**68.20**	**220.3635**		
50	100	125.2349	45.80	92.5940		
75	100	168.9605	50.60	99.6497		
100	100	217.0383	53.60	129.4235		
k	$\mu_{u,1}$ (m/s)	$\mu_{\omega,1}$ (rad/s)	$\mu_{u,2}$ (m/s)	$\mu_{\omega,2}$ (rad/s)	$t_f\mu_{u,2}$ (m)	$t_f\mu_{\omega,2}$ (rad)
1	73.2869	9.0937	0.0423	0.0016	126.7413	4.7963
5	73.7068	10.1675	0.2139	0.0081	130.4128	4.9310
15	**76.4238**	**17.1043**	**0.5888**	**0.0187**	**129.7534**	**4.1278**
50	103.6473	46.4832	2.0498	0.0531	189.7969	4.9211
75	127.2770	65.3029	2.2921	0.0555	228.4063	5.5290
100	156.9870	91.0888	2.1674	0.0521	280.5187	6.7436

- Collisions always remain above 60%.
- The time t_f to reach formation increases with $\bar{\alpha}$.
- All four input measures are inversely proportional to $\bar{\alpha}$. The maximum angular velocity is as high as 994 rad/s when $\bar{\alpha} = 0.1$. This would certainly require saturation.
- The lower the value of $\bar{\alpha}$, the more oscillatory the response. The most revolutions that a unicycle makes is 1.58 when $\bar{\alpha} = 0.1$.

The main conclusions from the simulation results in Table 9.6, obtained by varying k are:

- For $k < 15$ (the reference value), some simulations fail. Only 0.4% of simulations fail when k is chosen as low as 1.
- The drift is proportional to k.
- Collisions attain a minimum of 45% when $k = 50$.
- The time to reach formation attains a minimum of 92.59 s when $k = 50$.
- All four input measures are proportional to k.
- The most revolutions that a unicycle makes is 1.07 when $k = 100$.

The simulations reveal that we cannot choose $\bar{\alpha}$ and k much smaller than the reference values of $\bar{\alpha} = 5$ and $k = 15$ without running a risk of failure (i.e., formation may not be achieved). It is also important not to choose these values too large, especially $\bar{\alpha}$, as this can increase drift and/or t_f.

9.2.3 State-Dependent Undirected Graphs

In this section, we study how our control solution performs when the sensor graph is state dependent. We assume that each unicycle has a limited sensing radius. Consequently, the neighbors of a unicycle are defined as those that lie within its sensing circle determined by its sensing radius. We vary the sensing radius from 20m to 45m. The results are presented in Table 9.7. The column called *C/N* (%) represents the percentage of simulations that begin connected, *CF/C* (%) is the percentage of initially connected graphs that achieve formation, and *DF/D* (%) is the percentage of initially disconnected graphs that achieve formation. Naturally, the sensing radius has to be large enough to maintain connection when unicycles are in formation.

The main conclusions from the simulation results in Table 9.7 for state-dependent undirected graphs are:

- The proportion of initially connected undirected graphs goes from 5 to 95% as the sensing radius goes from 20 to 45 m.
- The proportion of initially connected undirected graphs that achieve formation goes from 59 to 98% as the sensing radius goes from 20 to 45 m.
- The proportion of initially disconnected undirected graphs that achieve formation goes from 11 to 70% as the sensing radius goes from 20 to 45 m.
- The sensing radius does not have much influence on the drift.

Table 9.7 Simulation results varying the sensing radius

rad.(m)	C/N (%)	CF/C (%)	DF/D (%)	μ_d (m)	Nc/N (%)	t_f (s)
20	5.80	58.62	11.25	74.5940	81.43	227.9798
25	22.20	68.47	24.68	72.4965	72.67	163.5336
30	45.40	79.30	44.69	70.5644	72.19	112.7208
35	72.20	89.20	45.32	71.9680	73.25	112.3022
40	86.20	94.90	52.17	70.2627	71.46	100.8281
45	95.40	97.90	69.57	69.6455	65.42	101.9338

rad.	$\mu_{u,1}$ (m/s)	$\mu_{\omega,1}$ (rad/s)	$\mu_{u,2}$ (m/s)	$\mu_{\omega,2}$ (rad/s)	$t_f\mu_{u,2}$ (m)	$t_f\mu_{\omega,2}$ (rad)
20	65.9074	15.7912	0.7856	0.0312	179.0938	7.1223
25	76.2224	16.2460	1.0513	0.0402	171.9238	6.5793
30	85.0675	19.1963	1.2967	0.0482	146.1645	5.4345
35	94.0355	19.9240	1.2721	0.0442	142.8599	4.9615
40	101.7096	21.7516	1.3289	0.0436	133.9915	4.3981
45	107.6055	23.6756	1.3108	0.0427	133.6198	4.3515

- t_f is inversely proportional to the sensing radius due to increased connectivity as the sensing radius increases.
- The input measures are typically proportional to the sensing radius.
- The maximum number of revolutions that a unicycle makes is inversely proportional to the sensing radius.

We also notice that the control solution appears to be robust with respect to the disconnection of some agents. This is illustrated in Fig. 9.8 where a single agent's initial position is too far from the rest of the group to sense anyone. Despite this, the formation is still attained among the remaining agents.

9.2.4 Directed Graphs

We present simulation results for the directed ring-coupled graph in Fig. 9.9 as opposed to the undirected graph in Fig. 9.6. The main conclusions from the simulation results in Table 9.8 are:

- 99% of the simulations achieved formation.
- The drift and collision measures are not changed significantly compared to the reference case.
- For this particular digraph, t_f is 2.23 times the value of the reference case. This makes sense because the undirected graph has 2.4 times the number of edge connections. The input measures are also higher in the reference case for the same reason.

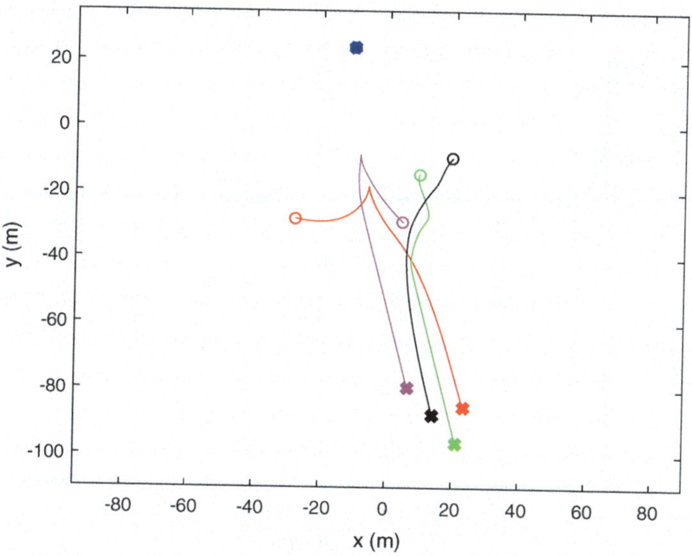

Fig. 9.8 Illustration of a formation where one agent is disconnected from the rest of the group and the rest of the group achieves formation. Initial positions are indicated with ∘, and final positions are indicated with ×

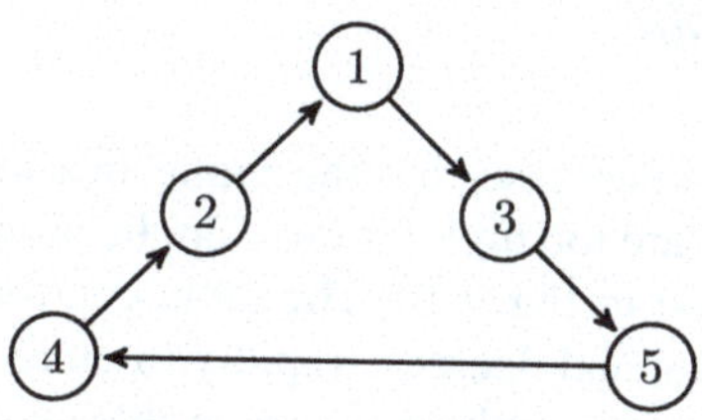

Fig. 9.9 Directed ring-coupled graph

Table 9.8 Simulation results for the digraph in Fig. 9.9

Graph	F/N (%)	μ_d (m)	Nc/N (%)	t_f (s)		
Fig. 9.6	**100**	**76.4330**	**68.20**	**220.3635**		
Fig. 9.9	99.00	79.7923	63.03	493.0773		

Graph	$\mu_{u,1}$ (m/s)	$\mu_{\omega,1}$ (rad/s)	$\mu_{u,2}$ (m/s)	$\mu_{\omega,2}$ (rad/s)	$t_f\mu_{u,2}$ (m)	$t_f\mu_{\omega,2}$ (rad)
Fig. 9.6	**76.4238**	**17.1043**	**0.5888**	**0.0187**	**129.7534**	**4.1278**
Fig. 9.9	38.0586	11.6481	0.2805	0.0102	138.2853	5.0263

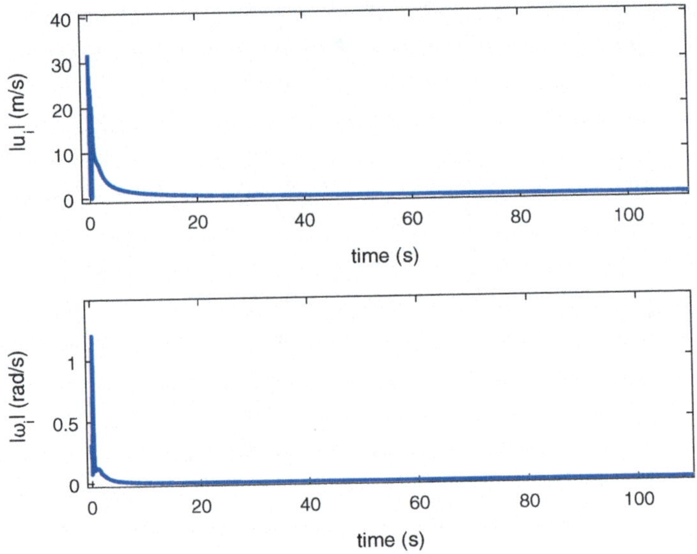

Fig. 9.10 Typical inputs magnitudes during simulation

9.2.5 *Input Saturation*

We have seen in the reference case that the input measures, $\mu_{u,1} = 76.4238$ m/s and $\mu_{\omega,1} = 17.1043$ rad/s, are too high for most applications. Typical profiles of the inputs u_i and ω_i are plotted in Fig. 9.10. The large control inputs are only present at the start of the simulation and decrease rapidly to more reasonable values as the formation is reached. This suggests that we may be able to saturate the inputs without affecting performance too much. In this section, we saturate the inputs so that $\mu_{u,1}$ and $\mu_{\omega,1}$ are bounded by the values corresponding to the first two columns in Table 9.9. Table 9.9 presents the simulation results where the bold quantities correspond to the reference case.

The profiles of the inputs u_i and ω_i for $\mu_{u,1} = 5$ m/s and $\mu_{\omega,1} = \pi/2$ rad/s are plotted in Fig. 9.11. The main conclusions from the simulation results in Table 9.9 for varying input saturation values are:

- The saturation did not lead to any significant formation failures (there was a single failure out of 500 tests for the case where $\mu_{u,1} = 15$ and $\mu_{\omega,1} = 3\pi/2$). This suggests that the solution is very robust to input saturation.
- The saturation did not lead to any significant change in drift and collisions.
- There was no significant increase in the formation time t_f unless $\mu_{u,1} < 1$ m/s and $\mu_{\omega,1} < \pi/4$ rad/s.
- There was no significant change in the mean control magnitudes $\mu_{u,2}$ and $\mu_{\omega,2}$ unless $\mu_{u,1} < 1$ m/s and $\mu_{\omega,1} < \pi/4$ rad/s, in which case the mean control magnitudes decrease.

Table 9.9 Simulation results with input saturation

$\mu_{u,1}$ (m/s)	$\mu_{\omega,1}$ (rad/s)	F/N (%)	μ_d (m)	Nc/N (%)	t_f (s)
0.2	$\pi/8$	100	80.5316	64.60	624.7621
1	$\pi/4$	100	78.3436	65.60	275.7020
5	$\pi/2$	100	76.2906	67.60	216.5130
10	π	100	78.2598	72.60	213.9683
15	$3\pi/2$	99.80	77.5945	69.34	216.5931
20	2π	100	75.3784	72.00	211.3089
76.4238	**17.1043**	**100**	**76.4330**	**68.20**	**220.3635**
$\mu_{u,1}$ (m/s)	$\mu_{\omega,1}$ (rad/s)	$\mu_{u,2}$ (m/s)	$\mu_{\omega,2}$ (rad/s)	$t_f\mu_{u,2}$ (m)	$t_f\mu_{\omega,2}$ (rad)
0.2	$\pi/8$	0.1687	0.01	105.4219	6.2186
1	$\pi/4$	0.4080	0.0199	112.4854	5.4835
5	$\pi/2$	0.5836	0.0230	126.3490	4.9879
10	π	0.5896	0.0209	126.1483	4.4680
15	$3\pi/2$	0.5959	0.0210	129.0616	4.5554
20	2π	0.6095	0.0208	128.7945	4.4015
76.4238	**17.1043**	**0.5888**	**0.0187**	**129.7534**	**4.1278**

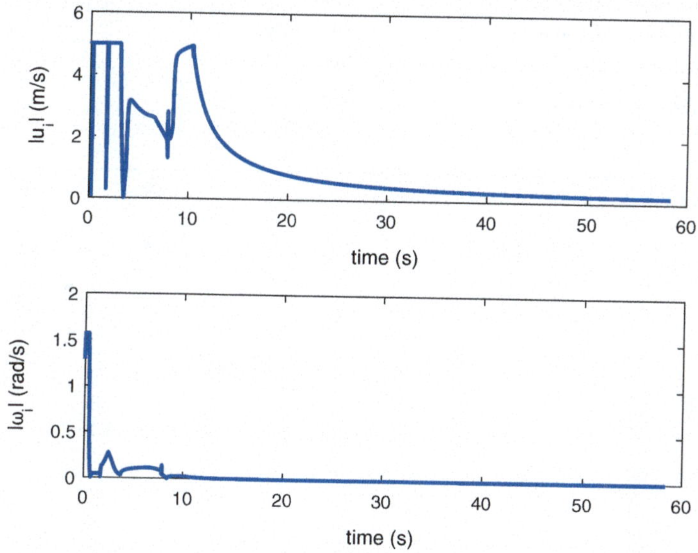

Fig. 9.11 Variation of the inputs magnitudes for $\mu_{u,1} = 5$ m/s and $\mu_{\omega,1} = \pi/2$ rad/s

9.2.6 *Disturbances and Sampling*

In this section, we present a simulation result including a number of disturbances as well as input sampling, which are not taken into account in the theoretical results. Due to slow simulation speed, only a single test is performed instead of 500 as in the other trials. Moreover, the inputs are saturated $\mu_{u,1} \leq 5$ m/s and $\mu_{\omega,1} \leq \pi/2$ rad/s. The simulation parameters are listed in Table 9.10. The disturbances are:

- an additive random noise with maximum magnitude of 0.25 m/s on the input u_i;
- an additive random noise with maximum magnitude of 0.25 rad/s on the input ω_i;
- an additive random noise with maximum magnitude of 0.25 rad on the quantity $g_i((\theta_{ij})_{j \in \mathcal{N}_i}, \eta)$ accounting for errors in measurements of relative headings;
- an additive random noise on $f_i(y_i^i)$, which account for measurement errors on the relative displacements of the vehicles. The direction of this vector has been

Table 9.10 Simulation parameters in Sect. 9.2.6

Quantity	Reference value
k	15
$\bar{\alpha}$	5
Fixed or state-dependent graphs	Fixed
Undirected or directed graph	Undirected
Input saturation	Yes
Sampling and disturbances	Yes

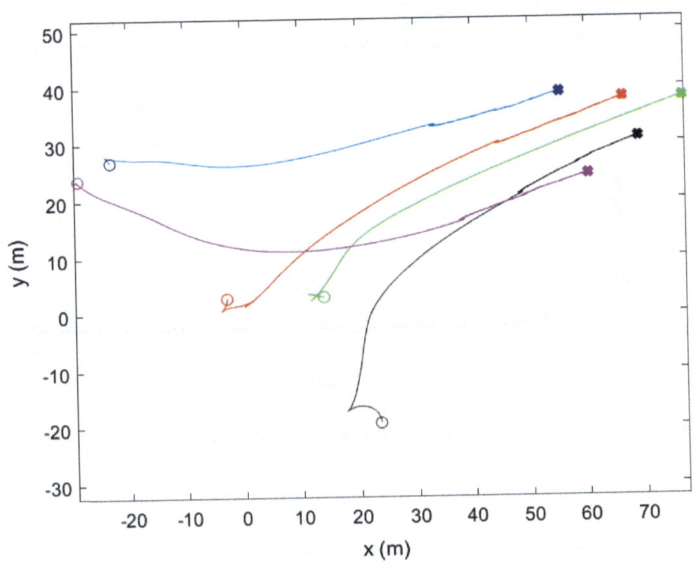

Fig. 9.12 Simulation result in the presence of disturbances and sampling rate 100 Hz

Table 9.11 Simulation results with perturbations and sampling

$\mu_{u,1}$ (m/s)	$\mu_{\omega,1}$ (rad/s)	$\mu_{u,2}$ (m/s)	$\mu_{\omega,2}$ (rad/s)
5	$\pi/2$	3.9174	1.2296

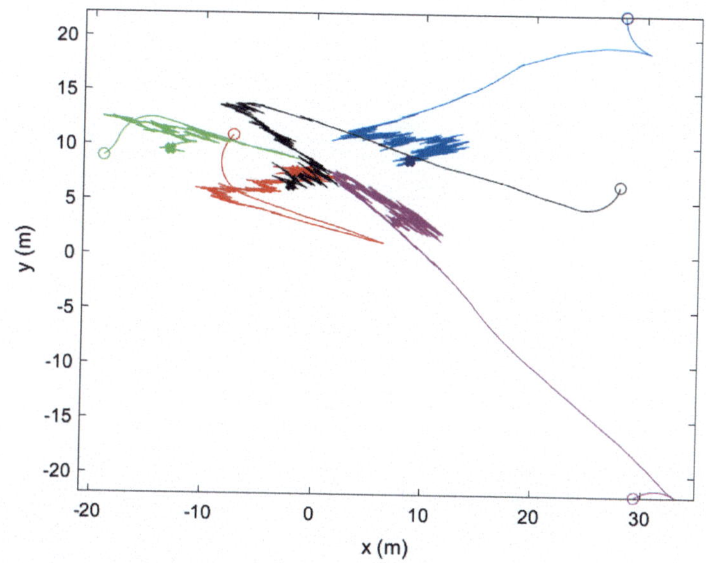

Fig. 9.13 Simulation result in the presence of disturbances and sampling rate 10 Hz

rotated within 0.25 rad, and the magnitude is scaled between 0.75 and 1.25 times the actual magnitude.

Each unicycle samples its input 100 times per second. The simulation results are presented in Fig. 9.12. The simulation results were run for 750 s, and the unicycles achieved the formation measure $\mu_f = 0.1123$ and no collisions. This shows that the control solution still performs well with perturbations and a sampling rate of 100 Hz. The formation drifted 74 m, and the input measures are reported in Table 9.11. In particular, the average input values $\mu_{u,2}$ and $\mu_{\omega,2}$ are significantly higher than in the reference case due to the disturbances. Reducing the sampling rate to 10 Hz no longer has acceptable performance as illustrated in Fig. 9.13 where the unicycles achieve the formation measure $\mu_f = 0.6455$ after 750 s and the unicycles's trajectories are erratic and nonsmooth.

Chapter 10
Bibliographical Notes

Abstract In this chapter, we discuss some literature on multi-agent coordination. The discussion is organized by research area. The references we discuss for each topic are by no means complete, but they serve as an introduction to the research area and contain the results that are most pertinent to the material of this book.

10.1 Literature on Multi-agent Coordination

10.1.1 Coordination Problems for Single and Double Integrators

The problem of rendezvous for networks of single and double integrators is well-established in the literature. See for example [3–6]. The majority of the literature on formation control considers single and double integrator models. A dominant approach for single integrator formation control is distance-based [7–9], where it is required that the distances between robots take on desired values. Often in this setting, the feedbacks are deduced from the gradient of a potential function whose minimum specifies the desired formation modulo roto-translations. This approach requires the sensing graph to be infinitesimally rigid, which is quite restrictive. Other approaches define formations in terms of relative angles between neighboring robots, instead of distances, [10, 11], or in terms of a complex Laplacian, [12–14]. In this latter case, formations are defined modulo scaling and roto-translations. Finally, formation flocking of double integrators is considered in [15], where the authors stabilize a formation and make sure that all robots in the formation achieve a common final velocity. See also [16].

A. Roza et al., *Distributed Coordination Theory for Robot Teams*, Lecture Notes in Control and Information Sciences 490, https://doi.org/10.1007/978-3-030-96087-2_10

10.1.2 Relative Equilibria for Kinematic Unicycles

The papers [17, 18] show that the only possible relative equilibria for a team of unicycles with local and distributed control laws correspond to either parallel (including stopping) or circular collective motions [2]. In fact, we will find local and distributed solutions to the problems studied in this work for unicycle rendezvous, stopping formations and parallel and circular formation flocking where the final collective motion is either parallel or circular. However, it is not possible to obtain formation flocking with any other type of collective motion, for example, around an oval shaped path using strictly local and distributed feedback. For kinematic unicycles in three dimensions, it is shown in [17, 19] that the only possible relative equilibria correspond to parallel, circular or helical formations. In [20], the authors propose distributed controllers to stabilize relative equilibria but do not specify a particular formation [21]. While kinematic unicycles in three dimensions are not studied explicitly in this work, we suspect that extension of the results from two dimensions to three dimensions would not cause any difficulties.

10.1.3 Kinematic Unicycle Rendezvous

In [22], the authors presented the first solution to this problem. The feedback in [22] is local and distributed and time-varying. In [23], the authors present a solution using a local and distributed, continuously differentiable and time-independent feedback. However, the result yields rendezvous only when the sensing graph is undirected and connected. The feedback in [23] makes the unicycles converge to a circular formation instead of rendezvous for some directed graphs containing a globally reachable node. In [24] the authors also consider undirected graphs and present a feedback to achieve rendezvous in finite time. In [25] both positions and attitudes of the unicycles are synchronized using a time-invariant distributed control. The authors assume an initially connected sensing graph. The controller that is implemented, however, is discontinuous and imposes excessive switching. In [26], a time-independent, local and distributed controller is presented. However, the authors make the assumption that whenever two vehicles get sufficiently close together they merge into a single vehicle, introducing a discontinuity in the control function. The same merging technique is used in [27] for cyclic and tree graphs where each unicycle keeps its neighboring vehicle within its windshield's field of view in order to maintain graph connectivity and achieve rendezvous. In [28, 29], distributed solutions are presented whereby the unicycles move toward the average position of their neighbors. However, a unicycle's feedback becomes undefined when it already lies at this average position which includes the case when the unicycles are at rendezvous. Finally, in [30], the authors solve the problem of practical rendezvous using a hybrid controller in which the unicycles converge to an arbitrarily small neighborhood of one another for undirected and connected graph topologies. The case of kinematic vehicles in three-space is investigated in [31–33]. The authors of [31, 32] consider the problem of full attitude and position synchronization, but assume fully actuated vehicles [21].

10.1.4 Flying Robot Attitude Synchronization and Rendezvous

The problem of attitude synchronization for flying robots is studied in [34–36]. The proposed controllers do not require measurements of the angular velocity, but they do require absolute attitude measurements. In [31], the authors use the energy shaping approach to design local and distributed controllers for attitude synchronization. The same approach is adopted in [37] to design two attitude synchronization controllers, both distributed. The first controller achieves almost-global synchronization for directed connected graphs. However, the controller design is based on distributed observers [38] and therefore requires auxiliary states to be communicated among neighboring vehicles. It also employs an angular velocity dissipation term that forces all vehicle angular velocities to zero in steady state. The second controller in [37] does not restrict the final angular velocities and does not require communication, but it requires an undirected sensing graph and guarantees only local convergence.

For the rendezvous problem of flying robots, in [39], the authors consider directed graphs containing a globally reachable node and develop an adaptive feedback that is not local and distributed. In [40, 41], the authors consider formation control for flying robots. However, again, the feedbacks are not local and distributed. In [41], the sensing graph is assumed to be undirected and communication among vehicles is required, while in [40], the graph is balanced, and it is assumed that each vehicle has access to the thrust input of its neighbors, therefore requiring once again communication between vehicles. Both approaches in [40, 41] use a two-stage backstepping methodology in which the first stage treats each vehicle as a point-mass system to which a desired thrust is assigned. A desired thrust direction is then extracted, and backstepping is used to design a rotational control such that vehicle rendezvous or formation control is achieved.

10.1.5 Formations of Kinematic Unicycles

A formation controller for single integrator robots can always be turned into a controller for kinematic unicycles if one considers a point at a positive distance d in front of each unicycle. As discussed in [42], these points behave like single integrators under an appropriate choice of feedback transformation and can be driven to a desired formation using the numerous techniques discussed earlier. However, although the points converge to a formation, the unicycles themselves do not. Choosing a small value of d reduces this error, but requires large control inputs. This results from the fact that kinematic unicycles have a nonholonomic constraint which is not present in the integrator model. For this reason, solutions for integrators do not adapt well into solutions for kinematic unicycles. We now discuss solutions in the literature designed specifically for kinematic unicycle formation control.

In [43], a discontinuous controller is presented that stabilizes formations with synchronized heading directions, but unicycles require a common sense of direction. The papers [22, 44, 45] discuss the feasibility of achieving various formations using local and distributed feedback [21]. In [22], time-dependent solutions are presented in each case. For general geometric patterns, unicycles require a common sense of direction. Similarly, the solution in [46] is time dependent and requires measurement of a common direction in addition to the velocity input of a neighboring unicycle, which can only be obtained if the unicycles can communicate these inputs with each other. In [47], a leader–follower approach is considered. The analysis transforms the unicycle model into a system of double integrators through dynamic feedback linearization. The desired formation is attained for digraphs containing a spanning tree, but each follower robot requires access to the acceleration of the leader through communication. In [48], each unicycle estimates its own position using dynamic extension, requiring communication among unicycles. The unicycles use these estimated states to attain the desired formation globally. The rotational control, however, is time dependent and oscillatory.

10.1.6 Kinematic Unicycle Formations with Final Collective Motions

Results in the literature for formation flocking remain quite limited. In [49], it is stated that "although some links between provably convergent formation control and flocking have been identified, the two behaviors have not been integrated into a single design." In [49], a solution is presented for parallel formation flocking that requires knowledge of a common frame (i.e., beacon) and uses a dynamic feedback linearization that requires communication of auxiliary states between agents. In [44], a local solution is presented for all-to-all undirected graphs. However, only stability is shown and not asymptotic stability. For circular formation flocking, [44, 50] present asymptotically stable solutions on a common circle with a focus on specific (M, N)-patterns. In [51], the authors present a controller with a repulsion function such that unicycles achieve equal spacing around a common circle assuming a jointly connected graph, while in [52], the spacing of the unicycles can be freely chosen beforehand. In these results, the feedbacks are local and distributed; however, the final formation is restricted to lie on a common circle and the stability results in [44, 50, 52] are local.

A far more studied problem in the literature is formation path following. In a common *virtual structure approach* [53–55] a (possibly virtual) leader moves along the desired path at a desired speed and the other unicycles converge to their corresponding place-holders with respect to the leader. In this approach, at least one unicycle must have access to the virtual leader state. This is trajectory tracking rather than path following, and the path is not invariant in this case. Another approach that is common in the literature [56–61], assigns a different path $\mathcal{C}_i(s_i)$ to each unicycle i where s_i parametrizes the displacement of unicycle i along its path. Formation path following is achieved whenever all unicycles lie on their corresponding paths and the path parameters for all unicycles achieve consensus. To achieve this, each unicycle invokes a path following controller in order to converge to its own path $\mathcal{C}_i(s_i)$ with a desired speed profile. In the transient behavior, the speed control is modified, slowing down or speeding up certain unicycles, in order to synchronize the s_i states. While this approach does not require a virtual leader, it does require communication of the quantities s_i between neighboring unicycles and each unicycle needs to compute its own path to be followed. In [62], unicycles converge to a common, compact, time-varying path with uniform spacing. Each unicycle stores, on-board, the state of an exosystem which must be communicated with neighboring unicycles. In [49], no virtual leader is required in the solution, but rather, the formation center of mass is driven to a desired path. However, each unicycle must know the location of the formation center of mass, requiring all-to-all sensing. Finally, in [63], the authors present a solution for hierarchical, leader–follower topologies. Unicycles approximately achieve formation as long as the path followed by the leader has sufficiently small curvature. However, the stabilizing control of a unicycle depends on the linear speed inputs of its neighbors.

There are several results in the literature that consider, specifically, formation circle following. Most consider motion along a common circle [44, 50, 64–66], while others, more in line with this book, allow unicycles to travel around a common center with different radii and correspondingly, with different speeds. This is the problem studied in [67] where the undirected sensing graph is assumed to be all-to-all and, by changing a gradient potential function in the control law, one can achieve either phase agreement or balancing. In [68], on the other hand, the graph is a ring and the spacing between neighboring unicycles are equal. Neither of these results, however, achieves arbitrary formations as in this book.

References

1. Roza, A., Maggiore, M., Scardovi, L.: A smooth distributed feedback for global rendezvous of unicycles. IEEE Trans. Control Netw. Syst. **5**(1), 640–652 (2018)
2. Roza, A., Maggiore, M., Scardovi, L.: A smooth distributed feedback for formation control of unicycles. IEEE Trans. Autom. Control **64**(12), 4998–5011 (2019)
3. Ren, W., Beard, R.: Consensus seeking in multiagent systems under dynamically changing interaction topologies. IEEE Trans. Autom. Control **50**(5), 655–661 (2005)
4. Moreau, L.: Stability of continuous-time distributed consensus algorithms. In: Proceedings of the 43rd IEEE Conference on Decision and Control, pp. 3998–4003 (2004)
5. Olfati-Saber, R., Murray, R.: Consensus problems in networks of agents with switching topology and time-delays. IEEE Trans. Autom. Control **49**(9), 1520–1533 (2004)

6. Ren, W., Atkins, E.: Distributed multi-vehicle coordinated control via local information exchange. Int. J. Robust Nonlinear Control **17**(10–11), 1002–1033 (2007)
7. Krick, L., Broucke, M., Francis, B.: Stabilisation of infinitesimally rigid formations of multi-robot networks. Int. J. Control **82**(3), 423–439 (2009)
8. Oh, K.-K., Ahn, H.-S.: Distance-based undirected formations of single-integrator and double-integrator modeled agents in n-dimensional space. Int. J. Robust Nonlinear Control **24**(12), 1809–1820 (2014)
9. Smith, S., Broucke, M., and Francis, B.: Stabilizing a multi-agent system to an equilateral polygon formation. In: Proceedings of the 17th International Symposium on Mathematical Theory of Networks and Systems, pp. 2415–2424 (2006)
10. Zhao, S., Lin, F., Peng, K., Chen, B., Lee, T.: Distributed control of angle-constrained cyclic formations using bearing-only measurements. Syst. Control Lett. **63**, 12–24 (2014)
11. Eren, T.: Formation shape control based on bearing rigidity. Int. J. Control **85**(9), 1361–1379 (2012)
12. Lin, Z., Ding, W., Yan, G., Yu, C., Giua, A.: Leader-follower formation via complex Laplacian. Automatica **49**(6), 1900–1906 (2013)
13. Lin, Z., Wang, L., Han, Z., Fu, M.: A graph Laplacian approach to coordinate-free formation stabilization for directed networks. IEEE Trans. Autom. Control **61**(5), 1269–1280 (2016)
14. Lin, Z., Wang, L., Han, Z., Fu, M.: Distributed formation control of multi-agent systems using complex Laplacian. IEEE Trans. Autom. Control **59**(7), 1765–1777 (2014)
15. Deghat, M., Anderson, B., Lin, Z.: Combined flocking and distance-based shape control of multi-agent formations. IEEE Trans. Autom. Control **61**(7), 1824–1837 (2016)
16. Tanner, H., Jadbabaie, A., and Pappas, G.: Stable flocking of mobile agents, part I: fixed topology. In: Proceedings of the 42nd IEEE Conference on Decision and Control, pp. 2010–2015 (2003)
17. Justh, E., Krishnaprasad, P.: Equilibria and steering laws for planar formations. Syst. Control Lett. **52**(1), 25–38 (2004)
18. Sarlette, A., Bonnabel, S., Sepulchre, R.: Coordinated motion design on Lie groups. IEEE Trans. Autom. Control **55**(5), 1047–1058 (2010)
19. Justh, E., Krishnaprasad, P.: Natural frames and interacting particles in three dimensions. In: Proceedings of the 44th IEEE Conference on Decision and Control, pp. 2841–2846 (2005)
20. Scardovi, L., Leonard, N., Sepulchre, R.: Stabilization of collective motion in three dimensions. Commun. Inf. Syst. **8**(3), 473–500 (2008)
21. Roza, A., Maggiore, M., Scardovi, L.: Local and distributed rendezvous of underactuated rigid bodies. IEEE Trans. Autom. Control **62**(8), 3835–3847 (2017)
22. Lin, Z., Francis, B., Maggiore, M.: Necessary and sufficient graphical conditions for formation control of unicycles. IEEE Trans. Autom. Control **50**(1), 121–127 (2005)
23. Zheng, R., Lin, Z., Cao, M.: Rendezvous of unicycles with continuous and time-invariant local feedback. In: Proceedings of the 18th IFAC World Congress, pp. 10044–10049, Milano, Italy (2011)
24. Zoghlami, N., Beji, L., Mlayeh, R., Abichou, A.: Finite-time consensus and stability of networked nonlinear systems. In: Proceedings of the 52nd IEEE Conference on Decision and Control, pp. 6604–6609 (2013)
25. Dimarogonas, D., Kyriakopoulos, K.: On the rendezvous problem for multiple nonholonomic agents. IEEE Trans. Autom. Control **52**(5), 916–922 (2007)
26. Zheng, R., Sun, D.: Rendezvous of unicycles: a bearings-only and perimeter shortening approach. Syst. Control Lett. **62**(5), 401–407 (2013)
27. Yu, J., LaValle, S., Liberzon, D.: Rendezvous without coordinates. IEEE Trans. Autom. Control **57**(2), 421–434 (2012)
28. Ajorlou, A., Asadi, M., Aghdam, A., Blouin, S.: Distributed consensus control of unicycle agents in the presence of external disturbances. Syst. Control Lett. **82**, 86–90 (2015)
29. Listmann, K., Masalawala, M., Adamy, J.: Consensus for formation control of nonholonomic mobile robots. In: Proceedings of the IEEE International Conference on Robotics and Automation, pp. 3886–3891 (2009)

30. Jafarian, M.: Ternary and hybrid controllers for the rendezvous of unicycles. In: Proceedings of the 54th IEEE Conference on Decision and Control, pp. 2353–2358 (2015)
31. Nair, S., Leonard, N.: Stable synchronization of rigid body networks. Netw. Heterogen. Media **2**(4), 595–624 (2007)
32. Dong, R., Geng, Z.: Consensus based formation control laws for systems on Lie groups. Syst. Control Lett. **62**(2), 104–111 (2013)
33. Hatanaka, T., Igarashi, Y., Fujita, M., Spong, M.: Passivity-based pose synchronization in three dimensions. IEEE Trans. Autom. Control **57**(2), 360–375 (2012)
34. Ren, W.: Distributed cooperative attitude synchronization and tracking for multiple rigid bodies. IEEE Trans. Control Syst. Technol. **18**(2), 383–392 (2010)
35. Abdessameud, A., Tayebi, A.: Attitude synchronization of a group of spacecraft without velocity measurements. IEEE Trans. Autom. Control **54**(11), 2642–2648 (2009)
36. Abdessameud, A., Tayebi, A.: Motion Coordination for VTOL Unmanned Aerial Vehicles. Springer (2013)
37. Sarlette, A., Sepulchre, R., Leonard, N.: Autonomous rigid body attitude synchronization. Automatica **45**, 572–577 (2009)
38. Scardovi, L., Sarlette, A., Sepulchre, R.: Synchronization and balancing on the N-torus. Syst. Control Lett. **56**(5), 335–341 (2007)
39. Wang, H.: Second-order consensus of networked thrust-propelled vehicles on directed graphs. IEEE Trans. Autom. Control **61**(1), 222–227 (2016)
40. Lee, D.: Distributed backstepping control of multiple thrust-propelled vehicles on a balanced graph. Automatica **48**(11), 2971–2977 (2012)
41. Abdessameud, A., Tayebi, A.: Formation control of VTOL unmanned aerial vehicles with communication delays. Automatica **47**(11), 2383–2394 (2011)
42. Francis, B., Maggiore, M.: Flocking and Rendezvous in Distributed Robotics. Springer (2016)
43. Dimarogonas, D., Kyriakopoulos, K.: A connection between formation infeasibility and velocity alignment in kinematic multi-agent systems. Automatica **44**(10), 2648–2654 (2008)
44. Sepulchre, R., Paley, D., Leonard, N.: Stabilization of planar collective motion: all-to-all communication. IEEE Trans. Autom. Control **52**(5), 811–824 (2007)
45. Tabuada, P., Pappas, G., Lima, P.: Motion feasibility of multi-agent formations. IEEE Trans. Robot. **21**(3), 387–392 (2005)
46. Jin, J., Gans, N.: Collision-free formation and heading consensus of nonholonomic robots as a pose regulation problem. Robot. Auton. Syst. **95**, 25–36 (2017)
47. Liu, T., Jiang, Z.-P.: Distributed formation control of nonholonomic mobile robots without global position measurements. Automatica **49**(2), 592–600 (2013)
48. Oh, K.-K., Ahn, H.-S.: Formation control of mobile agents based on distributed position estimation. IEEE Trans. Autom. Control **58**(3), 737–742 (2013)
49. Reyes, L., Tanner, H.: Flocking, formation control, and path following for a group of mobile robots. IEEE Trans. Control Syst. Technol. **23**(4), 1268–1282 (2015)
50. Sepulchre, R., Paley, D., Leonard, N.: Stabilization of planar collective motion with limited communication. IEEE Trans. Autom. Control **53**(3), 706–719 (2008)
51. Chen, Z., Zhang, H.-T.: No-beacon collective circular motion of jointly connected multi-agents. Automatica **47**(9), 1929–1937 (2011)
52. El-Hawwary, M., Maggiore, M.: Distributed circular formation stabilization for dynamic unicycles. IEEE Trans. Autom. Control **58**(1), 149–162 (2013)
53. Peng, Z., Wen, G., Rahmani, A., Yu, Y.: Distributed consensus-based formation control for multiple nonholonomic mobile robots with a specified reference trajectory. Int. J. Syst. Sci. **46**(8), 1447–1457 (2015)
54. Loria, A., Dasdemir, J., Jarquin, N.: Leader-follower formation and tracking control of mobile robots along straight paths. IEEE Trans. Control Syst. Technol. **24**(2), 727–732 (2016)
55. Sadowska, A., den Broek, T.V., Huijberts, H., van de Wouw, N., Kostić, D., Nijmeijer, H.: A virtual structure approach to formation control of unicycle mobile robots using mutual coupling. Int. J. Control **84**(11), 1886–1902 (2011)

56. Do, K., Pan, J.: Nonlinear formation control of unicycle-type mobile robots. Robot. Auton. Syst. **55**(3), 191–204 (2007)
57. Zhang, F., Leonard, N.: Coordinated patterns of unit speed particles on a closed curve. Syst. Control Lett. **56**(6), 397–407 (2007)
58. Ghabcheloo, R., Aguiar, A., Pascoal, A., Silvestre, C., Kaminer, I., Hespanha, J.: Coordinated path-following in the presence of communication losses and time delays. SIAM J. Control. Optim. **48**(1), 234–265 (2009)
59. Egerstedt, M., Hu, X.: Formation constrained multi-agent control. IEEE Trans. Robot. Autom. **17**(6), 947–951 (2001)
60. Chen, Y.-Y., Tian, Y.-P.: A curve extension design for coordinated path following control of unicycles along given convex loops. Int. J. Control **84**(10), 1729–1745 (2011)
61. Doosthoseini, A., Nielsen, C.: Coordinated path following for unicycles: a nested invariant sets approach. Automatica **60**, 17–29 (2015)
62. Brinón-Arranz, L., Seuret, A., Canudas-de Wit, C.: Cooperative control design for time-varying formations of multi-agent systems. IEEE Trans. Autom. Control **59**(8), 2283–2288 (2014)
63. Consolini, L., Morbidi, F., Prattichizzo, D., Tosques, M.: On a class of hierarchical formations of unicycles and their internal dynamics. IEEE Trans. Autom. Control **57**(4), 845–859 (2012)
64. Paley, D., Leonard, N., Sepulchre, R.: Stabilization of symmetric formations to motion around convex loops. Syst. Control Lett. **57**(3), 209–215 (2008)
65. Yu, X., Liu, L.: Distributed circular formation control of ring-networked nonholonomic vehicles. Automatica **68**, 92–99 (2016)
66. Yu, X., Xu, X., Liu, L., Feng, G.: Circular formation of networked dynamic unicycles by a distributed dynamic control law. Automatica **89**, 1–7 (2018)
67. Seyboth, G., Wu, J., Qin, J., Yu, C., Allgöwer, F.: Collective circular motion of unicycle type vehicles with nonidentical constant velocities. IEEE Trans. Control Netw. Syst. **1**(2), 167–176 (2014)
68. Zheng, R., Lin, Z., Fu, M., Sun, D.: Distributed control for uniform circumnavigation of ring-coupled unicycles. Automatica **53**, 23–29 (2015)

Appendix A
Notions of Stability Theory

In[1] this appendix we present concepts from stability theory used in this book.

A.1 Equilibrium Stability Theorems

We begin this chapter by reviewing a number of standard stability results taken from [2]. Here we assume the reader is familiar with the standard definitions of stability, attractivity, and asymptotic stability of an equilibrium found, e.g., in [2]. Consider the dynamical system

$$\dot{x} = f(x), \; x \in \mathcal{X}, \tag{A.1}$$

where $\mathcal{X} \subset \mathbb{R}^n$ is open and f is either locally Lipschitz on $\mathcal{X}$ or C^1. We assume that $0 \in \mathcal{X}$ and use the following notation. We denote by $\phi(t, x_0)$ the flow of f, i.e., the solution of (A.1) at time t with initial condition x_0. For a C^1 function $V(x)$, $L_f V(x) := (d/dx)V(x) \cdot f(x)$ denotes the Lie derivative of V along f.

Theorem A.1 (Lyapunov's direct method) *Let $x = 0$ be an equilibrium of* (A.1) *with $\mathcal{X} \subset \mathbb{R}^n$, and suppose there exists a C^1 function $V : D \to \mathbb{R}$, with $D \subset \mathcal{X}$ a domain containing* 0, *such that*

(i) V is positive definite at 0;
(ii) the Lie derivative $L_f V$ is negative definite at 0.

Then 0 *is an asymptotically stable equilibrium of* (A.1).

The global version of the Lyapunov Theorem is given below.

Theorem A.2 (Barbashin–Krasovskii) *Let $x = 0$ be an equilibrium of* (A.1) *with $\mathcal{X} = \mathbb{R}^n$, and suppose there exists a C^1 function $V : \mathbb{R}^n \to \mathbb{R}$ function such that V*

[1] ©2018 IEEE. Reprinted, with permission, from [1]. Reused Definition A.14 and Theorem A.5.

A. Roza et al., *Distributed Coordination Theory for Robot Teams*, Lecture Notes in Control and Information Sciences 490,
https://doi.org/10.1007/978-3-030-96087-2

is positive definite at 0*, radially unbounded, and* $L_f V$ *is negative definite at* 0*. Then,* $x = 0$ *is globally asymptotically stable.*

Theorem A.3 (Krasovskii–LaSalle Invariance Principle) *Let* $\Omega \subset D$ *be a compact positively invariant set. Let* $V : D \to \mathbb{R}$ *be a* C^1 *function such that for all* $x \in \Omega$*,* $L_f V$ *is negative semi-definite. Let* $E = \{x \in \Omega : L_f V = 0\}$ *and let* N *be the largest invariant subset of* E*. Then for all* $x_0 \in \Omega$*,* $\phi(t, x_0) \to N$ *as* $t \to \infty$*.*[2]

From Theorem A.3, it is also possible to obtain global results. In particular, if $D = \mathbb{R}^n$, every level set of V is compact, and if for all $x \in \mathbb{R}^n$, $L_f V$ is negative semi-definite, then for all $x_0 \in \mathbb{R}^n$, the solution $\phi(t, x_0)$ converges as $t \to \infty$ to N, the largest invariant set of $E = \{x \in \mathbb{R}^n : L_f V = 0\}$. The three theorems just stated have been presented with increasing generality with the Krasovskii–LaSalle Invariance Principle above being the most general as it does not require the function V to be positive definite and the attractor set is not necessarily a point.

The following proposition establishes global asymptotic stability of a neighborhood of an equilibrium.

Proposition A.1 *Let* $x = 0$ *be an equilibrium of* (A.1) *with* $\mathcal{X} = \mathbb{R}^n$*, and suppose there exists a* C^1 *function* $V : \mathbb{R}^n \to \mathbb{R}$ *such that* V *is positive definite at* 0*, has compact level sets, and* $L_f V$ *is negative outside the set* $V_c := \{x \in \mathbb{R}^n : V < c\}$*. Then,* V_c *is globally asymptotically stable.*

Proof Stability follows because $L_f V$ is negative outside the sub-level set $\bar{V}_c$ of the Lyapunov function V. For attractivity, consider any initial condition $x_0 \in \mathbb{R}^n$. There exists $\epsilon > c$ such that $x_0 \in V_\epsilon$. Since $L_f V$ is negative outside of V_c and $\bar{V}_\epsilon \backslash V_c$ is a compact set, $d := \max_{x \in \bar{V}_\epsilon \backslash V_c} L_f V(x) < 0$ is well-defined. Therefore, $\phi(t, x_0) \in V_c$ for all $t > (\epsilon - c)/d$ proving attractivity of V_c. □

Remark A.1 The three theorems and proposition stated above all require $V(x)$ to be a C^1 function. The proofs still hold, however, under the milder assumption that $V(x)$ and $L_f V(x)$ are both continuous.

A.2 Stability of Gradient Systems

We now apply the stability theorem just reviewed to the class of *gradient systems* defined as follows.

Definition A.1 A *gradient system* on the state space $\mathcal{X}$ is a differential equation of the form,

$$\dot{x} = -\nabla V(x), \quad x \in \mathcal{X}, \tag{A.2}$$

where $V : \mathcal{X} \to \mathbb{R}$ is C^2, and $\nabla V(x)$ is its gradient.

[2] The proof of the Krasovskii–LaSalle Invariance Principle (and in turn, the other theorems in this section) also apply when the state space is a smooth Riemannian manifold $(\mathcal{X}, g)$. See, for example, the proof in [2, Theorem 4.4].

The main feature of a gradient system is that the vector field $-\nabla V(x)$ is orthogonal to the level sets of the corresponding storage function V and point in the direction of steepest decent of V. The stationary points of the system correspond to the set where the gradient of V is zero. The function V becomes a natural candidate for a Lyapunov analysis. Taking the time derivative of V along system (A.2) yields,

$$\dot{V} = \frac{\partial V}{\partial x}\dot{x} = -\|\nabla V(x)\|^2 \leq 0.$$

Consider any isolated stationary point $\bar{x}$ which is a local minimum of V. Then $\bar{x}$ is an equilibrium for (A.2) and it immediately follows from the Lyapunov's direct method that $\bar{x}$ is asymptotically stable for (A.2). Gradient systems can also be analysed with the Krasovskii–LaSalle Invariance Principle. In particular, if V is defined globally on the domain $D = \mathbb{R}^n$ and its level sets are compact, then since the time derivative of V is negative semi-definite and equals zero only in the set $E = \{x \in \mathcal{X} : \nabla V(x) = 0\}$ of stationary points of V, we can conclude by the Krasovskii–LaSalle theorem that all solutions converge to the largest invariant set of stationary points of V. These observations are summarized in the following proposition.

Proposition A.2 (Gradient systems) *For the gradient system* (A.2), *where* $V : \mathcal{X} \to \mathbb{R}$ *is* C^2, *assume that all stationary points of* V *are isolated. Then, each stationary point of* V *that is also a local minimum of* V *is an asymptotically stable equilibrium of* (A.2). *Moreover, if* V *is proper, then all solutions of* (A.2) *are defined for all* $t \geq 0$ *and asymptotically converge to the set of stationary points of* V, $E = \{x \in \mathcal{X} : \nabla V(x) = 0\}$.

A.3 Stability of Homogeneous Systems

Now we turn our attention to the class of *homogeneous systems*, defined next.

Definition A.2 Let $\mathcal{X}, \mathcal{Y}$ be finite-dimensional real vector spaces and let Ω be a set. A function $f : \mathcal{X} \to \mathcal{Y}$ is *homogeneous of degree* $r \geq 0$ if, for all $\lambda > 0$ and for all $x \in \mathcal{X}$, $f(\lambda x) = \lambda^r f(x)$. A function $f : \mathcal{X} \times \Omega \to \mathcal{Y}$, $(x, y) \mapsto f(x, y)$, is *homogeneous of degree r with respect to x* if for all $\lambda > 0$ and for all $(x, y) \in \mathcal{X} \times \Omega$, $f(\lambda x, y) = \lambda^r f(x, y)$.

By this definition, a homogeneous function of degree r is one that scales by powers of r moving outward along rays going through the origin. Examples of homogeneous functions are

- The function $f(x, y) = \text{atan2}(y, x)$ is homogeneous of degree zero. Notice that f is undefined at $(x, y) = (0, 0)$.
- Linear functions are homogeneous of degree one directly by the scalar multiplication property $f(\lambda x) = \lambda x$ for any $\lambda > 0$.

- The real function $f(x, y) = xy$ is homogeneous of degree two since $f(\lambda x, \lambda y) = \lambda^2 xy$ for any scalar $\lambda > 0$.
- The real function $f(x, y) = x^2 y$ is homogeneous of degree two with respect to x and one with respect to y.

If the function $f(x)$ is homogeneous of degree r and $g(x)$ is homogeneous of degree s then the product $h(x)$ is homogeneous of degree $r + s$ since

$$h(\lambda x) = f(\lambda x)g(\lambda x) = \lambda^{r+s} f(x)g(x) = \lambda^{r+s} h(x).$$

Similarly, $h(x) = f(x)/g(x)$ is homogeneous of degree $r - s$.

Next, we present a number of useful properties of homogeneous functions.

Proposition A.3 *For a compact set Ω, let $f : \mathbb{R}^n \backslash \{0\} \times \Omega \to \mathbb{R}$, $(x, y) \mapsto f(x, y)$ be a continuous function, homogeneous of degree zero with respect to x. Then f achieves a maximum on $\mathbb{R}^n \backslash \{0\} \times \Omega$.*

Proof Since $f(x, y)$ is homogeneous of degree zero with respect to x, it follows that for all $x \neq 0$, $f(x, y) = \|x\|^0 f(x/\|x\|, y) = f(x/\|x\|, y)$. Since f is continuous and $(x/\|x\|, y)$ lie on compact sets, $f(x/\|x\|, y)$ has a maximum value on the domain $(\mathbb{R}^n \backslash \{0\}) \times \Omega$. □

Now consider the dynamical system

$$\Sigma : \dot{x} = f(x), \; x \in \mathcal{X}, \tag{A.3}$$

where $\mathcal{X}$ is a finite-dimensional vector space. We say that Σ in (A.3) is a *homogeneous system of degree* r if f is a homogeneous function of degree r.

Proposition A.4 below shows that if we have a homogeneous system and a positive definite homogeneous Lyapunov function V, then to prove global asymptotic stability of the origin 0 it is enough to show negative definiteness of $L_f V$ on a compact set rather than the entire state-space. This is beneficial because compact sets enjoy several useful properties. For example, the fact that continuous real functions attain a maximum on compact sets.

Proposition A.4 *Consider the homogeneous system of degree $r > 0$ in* (A.3)*, and assume that $\mathcal{X} = \mathbb{R}^n$. Suppose that there exists a continuous positive definite function $V : \mathbb{R}^n \to \mathbb{R}$, homogeneous of degree $s \geq 1$, such that $V^{-1}(0) = \{0\}$ and suppose that $L_f V(x)$ is continuous and satisfies $L_f V(\phi) < 0$ for all $\phi \in \{x \in \mathbb{R}^n : \|x\| = 1\} \cong \mathbb{S}^{n-1}$. Then the set $\{0\}$ is GAS.*

Proof Since $V(x)$ is homogeneous of degree $s \geq 1$ for all $\lambda > 0$ and for all $x \in \mathcal{X}$, $V(\lambda x) = \lambda^s V(x)$. It follows from Euler's Theorem presented in Sect. 12.8 in [3] that the derivative $DV(x) := \partial V(x)/\partial x$ is homogeneous of degree $s - 1$. Moreover, the level sets of $V(x)$ are compact because V is homogeneous of degree $s \geq 1$ and continuous. One can write $x = \lambda \phi$ where $\lambda = \|x\|$ and $\phi \in \mathbb{S}^{n-1}$ is a unit vector. The Lie derivative $L_f V(x)$ therefore satisfies

$$L_f V(x) = DV(\lambda\phi)f(\lambda\phi),$$

which by the homogeneity properties of DV and f becomes,

$$L_f V(x) = \lambda^{r+(s-1)} DV(\phi)f(\phi) = \lambda^{r+(s-1)} L_f V(\phi),$$

where $r + (s-1) > 0$. By assumption, for all $\phi \in \mathbb{S}^{n-1}$, $L_f V(\phi) < 0$. Therefore $L_f V(x) \leq 0$ with equality if and only if $\lambda = \|x\| = 0$ and it follows that $L_f V(x)$ is negative semi-definite. All level sets of $V(x, \theta)$ are therefore positive invariant. By the Krasovskii–LaSalle Invariance Principle $\{0\}$ is globally attractive. Moreover stability follows because $V^{-1}(0) = \{0\}$. □

The proofs in Chap. 5 and, particularly, Chap. 6 use similar arguments to Proposition A.5 below. This proposition is presented primarily as an aid to understanding the logic in those proofs.

Proposition A.5 *Consider the homogeneous system of degree $r > 0$ in* (A.3)*, and assume that $\mathcal{X} = \mathbb{R}^n$. Let $V : \mathcal{X} \to \mathbb{R}$ be a continuous positive definite function, homogeneous of degree $s \geq 2$. Let $g : \mathcal{X} \to \mathbb{R}$ be a continuous function, homogeneous of degree $s - 1$. Finally, let*

$$W(x) = \alpha\sqrt{V(x)} + g(x)$$

for $\alpha > 0$. Then there exists $\alpha^\star > 0$ such that for all $\alpha > \alpha^\star$, $W(x)$ is positive definite. Moreover, if $L_f W(x)$ is continuous and $L_f W(\theta) < 0$ for all $\theta \in \{x \in \mathbb{R}^n : \|x\| = 1\} \cong \mathbb{S}^{n-1}$ and for all $\alpha > \alpha^\star$, then $x = 0$ is globally asymptotically stable.

Proof The function $W(x)$ is continuous and homogeneous of degree $s - 1$ because both $\sqrt{V(x)}$ and $g(x)$ are such. First it is shown that there exists $\alpha^\star > 0$ such that for all $\alpha > \alpha^\star$, $W(x)$ is positive definite. For $x \neq 0$, $V(x) > 0$ and one can write

$$W(x) = \sqrt{V(x)}\left(\alpha + \frac{g(x)}{\sqrt{V(x)}}\right).$$

The function $g(x)/\sqrt{V(x)}$ is homogeneous of degree zero because the numerator and denominator are both homogeneous of degree $s - 1$ and therefore, by Proposition A.3, it attains a maximum over $\mathcal{X}\backslash\{0\}$. Therefore, choosing $\alpha^\star > \max_{x\in\mathcal{X}\backslash\{0\}} \|g(x)/\sqrt{V(x)}\|$ implies that $\beta := \alpha^\star + g(x)/\sqrt{V(x)} > 0$. Then, for all $x \neq 0$, $W(x) > \sqrt{V(x)}\beta > 0$ since $V(x)$ is positive definite and $W(0) = 0$ since $W(x)$ is homogeneous of degree $s - 1 > 0$. $W(x)$ is radially unbounded because it is both positive definite and homogeneous of degree $s - 1 > 0$. Moreover, if $L_f W(\theta) < 0$ for all $\theta \in \{x \in \mathbb{R}^n : \|x\| = 1\} \cong \mathbb{S}^{n-1}$ and for all $\alpha > \alpha^\star$, then it follows from Proposition A.4 that $x = 0$ is globally asymptotically stable. □

A.4 Exponential Instability of Equilibria

In this section we present a useful result, due to Freeman [4], concerning the domain of attraction of equilibria with unstable linearization. Consider again the dynamical system

$$\dot{x} = f(x), \; x \in \mathcal{X}, \tag{A.4}$$

where $\mathcal{X} \subset \mathbb{R}^n$ is open and f is C^1.

Definition A.3 An equilibrium $p \in \mathcal{X}$ of system (A.4) *exponentially unstable* if the differential df_p of f at p has at least one eigenvalue in the open right-half complex plane.

The definitions for exponential stability and instability of an equilibrium point can similarly be defined relative to a subset $\Gamma \subset \mathcal{X}$, as stated next.

Definition A.4 Consider the dynamical system Σ in (A.4), and let Γ be a closed embedded submanifold of $\mathcal{X}$ that is invariant for Σ. If $p \in \Gamma$ is an equilibrium of f, then p is *exponentially unstable relative to* Γ if p is exponentially unstable for $f|_\Gamma$, i.e., the differential at p of a coordinate representation of $f|_\Gamma$ has at least one eigenvalue in the open right-half complex plane.

Theorem A.4 below, due to Freeman [4], allows us to say when the domain of attraction of a set of exponentially unstable equilibria $\mathcal{A}$ is Lebesgue measure zero. We use this definition in Theorem A.4 in Sect. A.6, where we develop a novel reduction theorem which plays a key role in Chap. 8. In what follows, $D(\mathcal{A})$ denotes the domain of attraction of a set $\mathcal{A}$. This is formally defined in Definition A.6 below.

Theorem A.4 (Proposition 1 in [4]) *Suppose $\mathcal{A}$ is a set of equilibria for system Σ in* (A.4), *each equilibrium in $\mathcal{A}$ is exponentially unstable, and*

$$D(\mathcal{A}) = \bigcup_{z \in \mathcal{A}} D(\{z\}). \tag{A.5}$$

Then $D(\mathcal{A})$ is meager[3] *and has zero measure.*

Condition (A.5) of Theorem A.4 is satisfied if and only if a solution converging to the set $\mathcal{A}$, necessarily converges to one of the exponentially unstable equilibria in the set $\mathcal{A}$.

A.5 Stability of Sets: Definitions

All the distributed control problems discussed in this book are formulated in terms of stability of certain sets. In this section we present a number of the set stability

[3] A set is *meager* if it can be expressed as the union of countably many nowhere dense subsets.

notions. For a detailed discussion, the reader can refer to [5]. We begin with a list of the stability definitions that we present in this section. For convenience, we use abbreviations to refer to various stability concepts (the abbreviations are summarized at the beginning of this book).

- Set stability (Definition A.5).
- Domain of attraction (Definition A.6).
- Attractivity and asymptotic stability (Definition A.7).
- Almost global attractivity and asymptotic stability (Definition A.10).
- Semiglobal attractivity and asymptotic stability (Definition A.11).
- Almost semiglobal attractivity and asymptotic stability (Definition A.12).
- Global practical stability (Definition A.13).
- Stability of a set relative to another set, and local stability of a set near another set (Definition A.14).

Let $\mathcal{X} \subset \mathbb{R}^n$ be open[4]. If $d : \mathcal{X} \times \mathcal{X} \to [0, \infty)$ is the distance function between two points in $\mathcal{X}$ and $\Gamma \subset \mathcal{X}$ is a closed subset of $\mathcal{X}$, then we denote by $\|x\|_\Gamma := \inf_{\psi \in \Gamma} d(x, \psi)$ the point-to-set distance of $x \in \mathcal{X}$ to Γ. If $\varepsilon > 0$, we let $B_\varepsilon(\Gamma) := \{x \in \mathcal{X} : \|x\|_\Gamma < \varepsilon\}$ and by $\mathcal{N}(\Gamma)$ we denote a neighborhood of Γ in $\mathcal{X}$.

In the discussion that follows, consider a smooth dynamical system

$$\Sigma : \dot{x} = f(x), \tag{A.6}$$

with state space $\mathcal{X}$ and solutions defined for all time $t \geq 0$, and let $\phi(t, x_0)$ denote the solution at time t with initial condition $x(0) = x_0$. A closed set $\Gamma \subset \mathcal{X}$ is said to be positively invariant for Σ if for all $x_0 \in \Gamma$ and all $t > 0$, $\phi(t, x_0) \in \Gamma$.

Roughly, a closed subset $\Gamma \subset \mathcal{X}$ for system (A.6) is *stable* if solutions starting close to Γ remain close to Γ for all time. The precise definition for stability is given in Definition A.5 and is illustrated on the left hand side of Fig. A.1.

Definition A.5 The closed set $\Gamma \subset \mathcal{X}$ is *stable* for Σ if for any $\varepsilon > 0$, there exists a neighborhood $\mathcal{N}(\Gamma) \subset \mathcal{X}$ such that, for all $x_0 \in \mathcal{N}(\Gamma)$, $\phi(t, x_0) \in B_\varepsilon(\Gamma)$, for all $t > 0$.

Stability is a key factor to ensure robustness in engineering applications. For example, consider the problem of rendezvous where Γ represents the rendezvous set in which the positions of all agents coincide. One would expect from a good control solution that if initial robot positions are close to rendezvous, i.e., if the initial state is in a small neighborhood of Γ then the corresponding closed-loop solution should not diverge too much from Γ. If a set is not stable then it is called *unstable*.

The *domain of attraction* of a closed set $\Gamma \subset \mathcal{X}$ for system (A.6), denoted $D(\Gamma)$, is the set of initial conditions in the state space $\mathcal{X}$ from which solutions converge to the set Γ as time approaches infinity. The notion of domain of attraction is formally stated in Definition A.6.

---

[4] One can consider, more generally, a complete Riemannian manifold $(\mathcal{X}, g)$ with associated Riemannian distance function $d : \mathcal{X} \times \mathcal{X} \to [0, \infty)$ on $\mathcal{X}$.

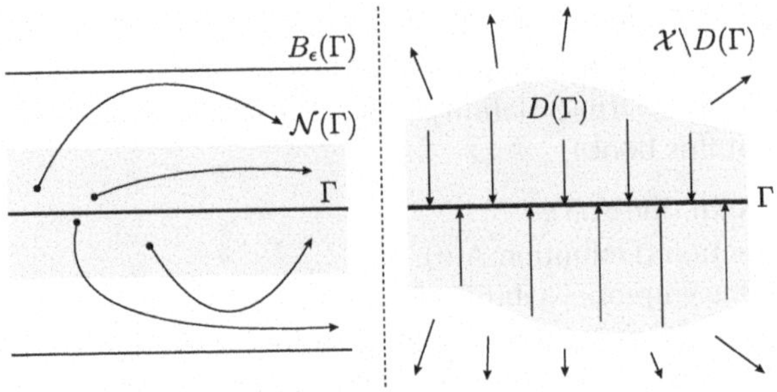

Fig. A.1 Illustration on the left shows stability of Γ - solutions with initial conditions in $\mathcal{N}(\Gamma)$ remain inside $B_\varepsilon(\Gamma)$ for all time. Illustration on the right shows attractivity of Γ with domain of attraction $D(\Gamma)$ - solutions with initial conditions in $D(\Gamma)$ converge to Γ

Definition A.6 The *domain of attraction* of the closed set $\Gamma \subset \mathcal{X}$ for system Σ is the set $D(\Gamma) = \{x_0 \in \mathcal{X} : \lim_{t\to\infty} \|\phi(t, x_0)\|_\Gamma = 0\}$.

Definition A.7 The closed set $\Gamma \subset \mathcal{X}$ is *(locally) attractive* for Σ if $D(\Gamma)$ is a neighborhood of Γ; Γ is *globally attractive* for Σ if $D(\Gamma) = \mathcal{X}$. Moreover, Γ is *locally asymptotically stable* (LAS) for Σ if Γ is stable and locally attractive; Γ is *globally asymptotically stable* (GAS) for Σ if Γ is stable and globally attractive.

While it is desirable to have global results, asking for global asymptotic stability is often too strong in many applications. This may be because this property is simply too difficult to prove or, more fundamentally, it may not even be possible to design continuous time-invariant control inputs to make a desired closed subset $\Gamma \subset \mathcal{X}$ globally asymptotically stable due to a topological obstruction. In fact, this issue is quite common. Take for example Theorem 1 in [6], which implies that any smooth time-invariant vector field on a compact manifold without boundary cannot have any equilibrium that is globally asymptotically stable. Consequently, it is impossible to globally asymptotically stabilize an equilibrium point in $SO(2)$ and $SO(3)$ via continuous time-invariant feedback. For this reason, we will define a slightly weaker form of global asymptotic stability called *almost global asymptotic stability* in which the domain of attraction is not global, but rather, contains almost every point in the state-space.

Before giving a formal definition of almost global asymptotic stability, we need to understand what a set of Lebesgue measure zero is. We will start by defining Lebesgue measure zero sets in $\mathbb{R}^n$ and then extend this definition to smooth manifolds. Intuitively, a set of Lebesgue measure zero is negligible in the sense that it occupies no volume in the state space and the probability of randomly choosing a point in this set is therefore zero. Define an open cube in $\mathbb{R}^n$ as the product of open intervals $U = (a_1, b_1) \times (a_2, b_2) \cdots \times (a_n, b_n)$ where $a_i, b_i \in \mathbb{R}$ and $a_i < b_i$ for all $i \in \mathbf{n}$. Denote the volume of this cube by the product of the intervals $v(U) = (b_1 - a_1)(b_2 - a_2)\ldots(b_n - a_n)$. A set of Lebesgue measure zero in $\mathbb{R}^n$ is one that can be covered by a collection of cubes occupying an arbitrarily small volume.

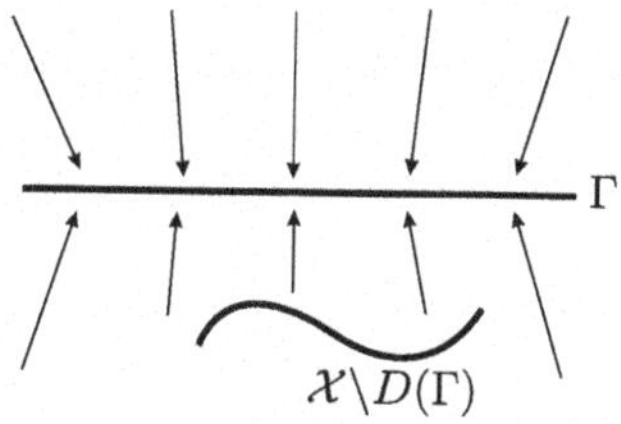

Fig. A.2 Illustration of almost global attractivity of the set Γ. The set $\mathcal{X}\backslash D(\Gamma)$ has Lebesgue measure zero

Definition A.8 A subset $\Gamma \subset \mathbb{R}^n$ has zero measure if for every $\epsilon > 0$, there exist open cubes $U_1, U_2, \ldots$ such that $\Gamma \subset \bigcup_{i=1}^{\infty} U_i$ and

$$\sum_{i=1}^{\infty} v(U_i) < \epsilon.$$

In the case of a smooth manifold M, a set $\Gamma \subset M$ has Lebesgue measure zero in M if it has zero measure under every smooth coordinate chart on M [7]. Moreover, by Proposition 6.8 in [7] $M\backslash\Gamma$ is dense in M. This means that the closure of $M\backslash\Gamma$ satisfies $\overline{M\backslash\Gamma} = M$.

Definition A.9 If M is a smooth n-dimensional manifold, a subset $\Gamma \subset M$ has *zero measure in* M if for every smooth coordinate chart (U, φ) for M, the subset $\varphi(\Gamma \cap U) \subset \mathbb{R}^n$ has measure zero.

We are now ready to define almost global stability properties.

Definition A.10 The closed set $\Gamma \subset \mathcal{X}$ is *almost globally attractive* for Σ if $\mathcal{X}\backslash D(\Gamma)$ has Lebesgue measure zero. The set Γ is *almost globally asymptotically stable* (AGAS) for Σ if Γ is stable and almost globally attractive. △

The notion of almost global attractivity is illustrated in Fig. A.2. Next we give two final definitions of attractivity and asymptotic stability which require a high gain control $k \in \mathbb{R}$. Consider the dynamical system

$$\Sigma(k) : \dot{x} = f(x, k) \tag{A.7}$$

and let $\phi_k(t, x_0)$ denote the solution at time t with initial condition $x(0) = x_0$. For system (A.7) the domain of attraction will depend on the parameter k in general and is denoted by $D_k(\Gamma)$.

Definition A.11 The closed set $\Gamma \subset \mathcal{X}$ is *semiglobally attractive with high-gain parameter* k for $\Sigma(k)$ if for each compact set K satisfying $\Gamma \subset K \subset \mathcal{X}$, there exists $k^\star > 0$ such that for all $k > k^\star$, Γ is attractive for $\Sigma(k)$ and $K \subset D_k(\Gamma)$. The set Γ is *semiglobal asymptotically stable* (SGAS) for $\Sigma(k)$ if Γ is stable and semiglobally attractive. △

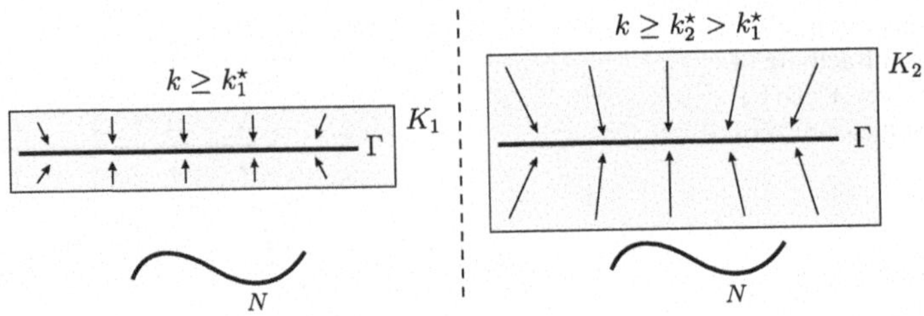

Fig. A.3 Illustration of almost semiglobal attractivity of the set Γ. For compact sets $K_1 \subset K_2 \subset \mathcal{X}\backslash N$, that do not contain the set N of Lebesgue measure zero, for all $k \geq k_1^\star$ (left) and $k \geq k_2^\star > k_1^\star$ (right), solutions with initial conditions in the sets K_1 and K_2 respectively, converge to Γ. The domain of attraction of Γ approaches full measure as the high gain k is increased

Definition A.12 A closed subset $\Gamma \subset \mathcal{X}$ is *almost semiglobally attractive with high-gain parameter* k for $\Sigma(k)$ if there exists a set $N \subset \mathcal{X}\backslash\Gamma$ of Lebesgue measure zero such that for each compact subset K satisfying $\Gamma \subset K \subset (\mathcal{X}\backslash N)$, there exists $k^\star > 0$ such that for all $k > k^\star$, Γ is attractive for $\Sigma(k)$ and $K \subset D_k(\Gamma)$. The set Γ is *almost semiglobally asymptotically stable* (ASGAS) for $\Sigma(k)$ if Γ is stable and almost semiglobally attractive. See Fig. A.3. △

The difference between global asymptotic stability and semiglobal asymptotic stability of a closed subset $\Gamma \subset \mathcal{X}$ is that with the former, solutions converge to the set Γ from all initial conditions, while with the latter, the domain of attraction can be made arbitrarily large within the state-space by increasing the control gain k. The difference between almost global asymptotic stability and almost semiglobal asymptotic stability is that with the former, solutions converge to the set Γ with domain of attraction $D(\Gamma)$ of full measure, while with the latter, the domain of attraction approaches full measure with increasing control gain k.

In place of asymptotic stability, one can also consider the weaker notion of *practical stability*.

Definition A.13 The closed set $\Gamma \subset \mathcal{X}$ is *globally practically stable* (GPS) for $\Sigma(k)$ if for any $\varepsilon > 0$, there exists $k^\star > 0$ such that for all $k > k^\star$, $B_\varepsilon(\Gamma)$ has a subset containing Γ which is globally asymptotically stable for $\Sigma(k)$. △

An illustration of global practical stability of the set Γ is shown in Fig. A.4. Notice the duality between the concepts of semiglobal stability and global practical stability. If Γ is asymptotically stable then it is necessarily practically stable without the requirement of a high gain parameter k. The purpose of considering practical stability instead of asymptotic stability is that in most engineering applications, it is sufficient that solutions converge to an arbitrarily small neighborhood $B_\varepsilon(\Gamma)$ of Γ, and not Γ itself. The advantage of considering notions of practical stability of a subset is that they are often significantly easier to prove. The downside to practical stability, however, is that it typically requires high gain. If the neighborhood $B_\varepsilon(\Gamma)$ is to be very small, then the gain will typically be large.

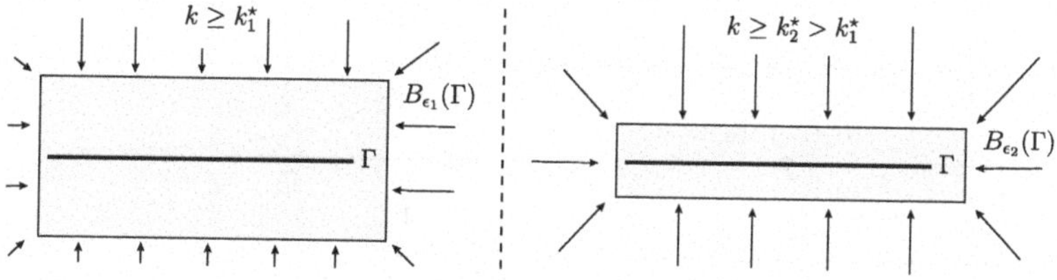

Fig. A.4 Illustration of global practical stability of the set Γ. For $\epsilon_1 > \epsilon_2$, for any $k \geq k_1^\star$ (left) and $k \geq k_2^\star > k_1^\star$ (right), for any initial conditions in $\mathcal{X}$, solutions converge to the neighborhoods $B_{\epsilon_1}(\Gamma)$ and $B_{\epsilon_2}(\Gamma)$ respectively

Now we generalize some of the definitions presented earlier. The following definitions are taken from [8].

Definition A.14 Let $\Gamma_1 \subset \Gamma_2$ be two subsets of $\mathcal{X}$ that are positively invariant for Σ. Assume that Γ_1 is compact and Γ_2 is closed.

- Γ_1 is *stable relative to* Γ_2 for Σ if, for any $\epsilon > 0$, there exists a neighborhood $\mathcal{N}(\Gamma_1)$ such that, $\phi(\mathbb{R}^+, \mathcal{N}(\Gamma_1) \cap \Gamma_2) \subset B_\epsilon(\Gamma_1)$. The notions of relative set attractivity, and asymptotic and practical stability are obtained analogously by restricting initial conditions to lie in Γ_2.
- Γ_2 is *locally stable near* Γ_1 if for all $x \in \Gamma_1$, for all $c > 0$ and all $\epsilon > 0$, there exists $\delta > 0$ such that for all $x_0 \in B_\delta(\Gamma_1)$ and all $t^\star > 0$, if $\phi([0, t^\star], x_0) \subset B_c(x)$ then $\phi([0, t^\star], x_0) \subset B_\epsilon(\Gamma_2)$.
- Γ_2 is *locally attractive near* Γ_1 if there exists a neighbourhood $\mathcal{N}(\Gamma_1)$ such that, for all $x_0 \in \mathcal{N}(\Gamma_1)$, $\|\phi(t, x_0)\|_{\Gamma_2} \to 0$ as $t \to \infty$.

The definitions for local stability of Γ_2 near Γ_1 and local attractivity of Γ_2 near Γ_1 are illustrated in Fig. A.5 adapted from [8]. Roughly, local stability of Γ_2 near Γ_1 means that for each ball $B_c(x)$, with $x \in \Gamma_1$, trajectories originating in $B_c(x)$ sufficiently close to Γ_1 cannot travel far away from Γ_2 before exiting $B_c(x)$. Local attractivity of Γ_2 near Γ_1, on the other hand, means that all initial conditions beginning in a neighborhood of Γ_1 converge to Γ_2 as $t \to \infty$. It can be seen immediately that the following implications hold:

- Γ_2 is stable $\implies$ Γ_2 is locally stable near Γ_1;
- Γ_1 is stable $\implies$ Γ_2 is locally stable near Γ_1;
- Γ_2 is attractive $\implies$ Γ_2 is locally attractive near Γ_1.

A.6 Stability of Sets: Reduction Theorems

We now present a powerful tool for analyzing the stability of sets. This tool is called *reduction* and allows one to break down the stability analysis of a set into simpler subproblems. The typical setup of the reduction problem is this. Suppose we are

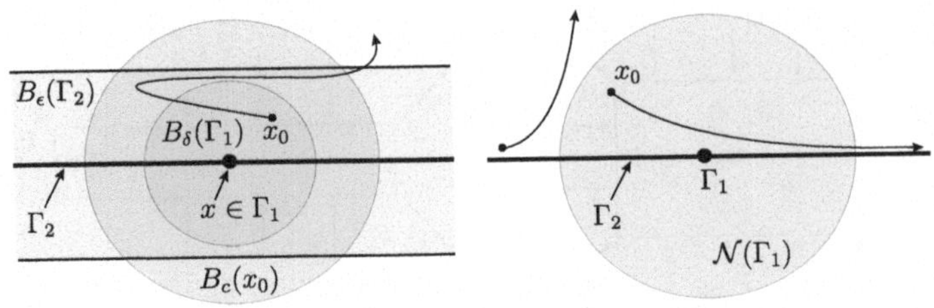

Fig. A.5 Illustration of local stability of Γ_2 near Γ_1 (left) and local attractivity of Γ_2 near Γ_1 (right)

given two sets $\Gamma_1 \subset \Gamma_2 \subset \mathcal{X}$, that are both positively invariant for system Σ, and suppose Γ_1 is, for instance, LAS relative to Γ_2. The reduction problem is to find conditions under which Γ_1 is LAS relative to $\mathcal{X}$. One may replace the property LAS by other properties, such as GAS, attractivity, AGAS, and so on. Here, we present a reduction theorem for LAS and GAS taken from [8, 9], and a novel reduction theorem for AGAS. The reader is referred to [8, 10] for a more detailed discussion of the reduction problem and its implications.

Theorem A.5 (Reduction Theorem *[8, 9]*) *Let Γ_1 and Γ_2, $\Gamma_1 \subset \Gamma_2 \subset \mathcal{X}$, be two closed sets that are positively invariant for Σ, and suppose Γ_1 is compact. Consider the following assumptions:*

(i) Γ_1 *is LAS relative to* Γ_2*;*
(i)′ Γ_1 *is GAS relative to* Γ_2*;*
(ii) Γ_2 *is locally stable near* Γ_1*;*
(iii) Γ_2 *is locally attractive near* Γ_1*;*
(iii)′ Γ_2 *is globally attractive;*
(iv) all trajectories of Σ *are bounded.*

Then, the following implications hold:

- *(i)* $\wedge$ *(ii)* $\implies \Gamma_1$ *is stable;*
- *(i)* $\wedge$ *(ii)* $\wedge$ *(iii)* $\iff \Gamma_1$ *is LAS;*
- *(i)′* $\wedge$ *(ii)* $\wedge$ *(iii)′* $\wedge$ *(iv)* $\iff \Gamma_1$ *is GAS.*

Note that one can replace condition (ii) with the stronger condition "Γ_2 is stable" and replace (iii) with the stronger condition "Γ_2 is asymptotically stable". Then it is clear conditions (ii) and (iii) together are satisfied when Γ_2 is asymptotically stable and (ii) and (iii)′ together are satisfied when Γ_2 is globally asymptotically stable.

The next result presents a novel reduction theorem for almost global asymptotic stability. This theorem has not been presented elsewhere.

Theorem A.6 *Let Γ_1 and Γ_2, $\Gamma_1 \subset \Gamma_2 \subset \mathcal{X}$, be two closed sets that are positively invariant for Σ, and suppose Γ_1 is compact. Consider the following assumptions:*

(i) Γ_2 *is an embedded submanifold of* $\mathcal{X}$ *which is globally asymptotically stable for* Σ*;*

(ii) Γ_1 *is globally attractive relative to* Γ_2 *and can be decomposed as a disjoint union* $\Gamma_1 = \mathcal{A} \bigsqcup \mathcal{K}$, *where* $\mathcal{A}$ *is a set of isolated equilibria which are exponentially unstable relative to* Γ_2, *and* $\mathcal{K}$ *is asymptotically stable relative to* Γ_2*;*

(iii) *all trajectories of* Σ *are bounded.*

Then $\mathcal{K}$ *is almost globally asymptotically stable.*

The proof of Theorem A.6 relies on the three lemmas presented next. The proof of Theorem A.6 is given after the presentation of the lemmas.

Lemma A.1 *Let* Γ *be a closed embedded submanifold of* $\mathcal{X}$ *which is invariant under a* C^1 *vector field* $f : \mathcal{X} \to T\mathcal{X}$. *Let* (U, φ) *be a smooth coordinate chart centred at* $p \in \Gamma$ *and let* $d\hat{f}_p$ *be the matrix representation of the differential* $df_p : T_p\mathcal{X} \to T_{f(p)}(T\mathcal{X})$ *in coordinates. If* p *is an equilibrium of* f, *then the subspace* $T_{\varphi(p)}\varphi(\Gamma \cap U)$ *of* $T_{\varphi(p)}\varphi(U)$ *is* $(d\hat{f}_p)$*-invariant.*

Proof Let $n = \dim \mathcal{X}$, $p \in \Gamma$ be such that $f(p) = 0$, and let $v_p \in T_p\Gamma$ be arbitrary with vector representation $\hat{v}_p \in T_{\varphi(p)}\varphi(\Gamma \cap U)$ in coordinates. We need to prove that $d\hat{f}_p\hat{v}_p \in T_{\varphi(p)}\varphi(\Gamma \cap U)$. Since Γ is embedded, there exists an open set $\bar{U} \subset U$ containing p and a smooth submersion $h : \bar{U} \to \mathbb{R}^{n-\dim\Gamma}$ such that $\Gamma \cap \bar{U} = h^{-1}(0)$. Let $I = (-1, 1)$ and $\sigma : I \to \mathcal{X}$ be a smooth regular curve whose image is contained in $\Gamma \cap \bar{U}$, and such that $\sigma(0) = p$ and $\dot{\sigma}(0) = v_p$. Since Γ is invariant under the flow of f, $f(\sigma(t)) \in T_{\sigma(t)}\Gamma$ for all $t \in I$ or, what is the same, $dh_{\sigma(t)}(f(\sigma(t))) = 0$ for all $t \in I$. Equivalently, $d\hat{h}_{\sigma(t)}\hat{f}(\sigma(t)) = 0$ for all $t \in I$ where $\hat{f}(\sigma(t))$ is the vector representation of $f(\sigma(t))$ and $d\hat{h}_{\sigma(t)}$ is the matrix representation of the differential $dh_{\sigma}(t) : T_{\sigma(t)}\bar{U} \to T_{h(\sigma(t))}\mathbb{R}^{n-\dim\Gamma}$ in coordinates. The derivative of $d\hat{h}_{\sigma(t)}\hat{f}(\sigma(t))$ with respect to t must be zero at $t = 0$,

$$\left.\frac{d}{dt}\right|_{t=0} \left[d\hat{h}_{\sigma(t)}\hat{f}(\sigma(t))\right] = 0,$$

or

$$\left(\left.\frac{d}{dt}\right|_{t=0} d\hat{h}_{\sigma(t)}\right)\hat{f}(\sigma(0)) + d\hat{h}_{\sigma(0)}\left.\frac{d}{dt}\right|_{t=0}\hat{f}(\sigma(t)) = 0.$$

Since $\hat{f}(\sigma(0)) = \hat{f}(p) = 0$, using the chain rule we get $d\hat{h}_{\sigma(0)}d\hat{f}_{\sigma(0)}\hat{v}_p = 0$, proving that $d\hat{f}_p\hat{v}_p \in T_p\Gamma$. □

Lemma A.2 *Let* Γ *be a closed embedded submanifold of* $\mathcal{X}$ *which is invariant under a* C^1 *vector field* $f : \mathcal{X} \to T\mathcal{X}$. *If* $p \in \Gamma$ *is an equilibrium of* f *which is exponentially unstable relative to* Γ, *then* p *is exponentially unstable relative to* $\mathcal{X}$.

Proof Let $n = \dim \mathcal{X}$ and $k = \dim \Gamma$. Since Γ is embedded, there exists a smooth coordinate chart (U, φ) centred at p, i.e., $\varphi(p) = 0$, such that $\varphi(\Gamma \cap U) = \{x \in \mathbb{R}^n : x^{k+1} = \cdots = x^n = 0\}$ where $x = (x^1, \ldots, x^k, x^{k+1}, \ldots, x^n)$ are local coordinates [7]. Since Γ is positively invariant, it follows that the restriction of the vector field $\hat{f}(x) = (\hat{f}_i(x))_{i\in\{1,\ldots,n\}}$ to $\Gamma \cap U$, represented in local coordinates, is given by $\hat{f}|_{\Gamma\cap U}(x^1, \ldots, x^k) = (\hat{f}_i(x^1, \ldots, x^k, 0, \ldots, 0))_{i\in\{1,\ldots,k\}}$. Since p is exponentially unstable relative to Γ, $d(\hat{f}|_{\Gamma\cap U})_{\varphi(p)}$ has at least one eigenvalue in the open right-half complex plane. To prove that p is exponentially unstable relative to $\mathcal{X}$, it needs to be shown that $d\hat{f}_{\varphi(p)}$ has at least one eigenvalue in the open right-half complex plane. It follows from Lemma A.1 that the tangent space $T_{\varphi(p)}\varphi(\Gamma \cap U)$ is $(d\hat{f}_{\varphi(p)})$-invariant in local coordinates and therefore $d\hat{f}_{\varphi(p)}$ has the upper triangular form

$$d\hat{f}_{\varphi(p)} = \begin{bmatrix} A_1 & A_2 \\ 0 & A_3 \end{bmatrix},$$

where

$$\begin{aligned} A_1 &= \left.\begin{bmatrix} \frac{d\hat{f}_1(x)}{dx^1} & \cdots & \frac{d\hat{f}_1(x)}{dx^k} \\ \vdots & & \vdots \\ \frac{d\hat{f}_k(x)}{dx^1} & \cdots & \frac{d\hat{f}_k(x)}{dx^k} \end{bmatrix}\right|_{x=0} \\ &= \left.\begin{bmatrix} \frac{d\hat{f}_1(x^1,\ldots,x^k,0,\ldots,0)}{dx^1} & \cdots & \frac{d\hat{f}_1(x^1,\ldots,x^k,0,\ldots,0)}{dx^k} \\ \vdots & & \vdots \\ \frac{d\hat{f}_k(x^1,\ldots,x^k,0,\ldots,0)}{dx^1} & \cdots & \frac{d\hat{f}_k(x^1,\ldots,x^k,0,\ldots,0)}{dx^k} \end{bmatrix}\right|_{(x^1,\ldots,x^k)=0} \\ &= d(\hat{f}|_{\Gamma\cap U})_{\varphi(p)} \end{aligned}$$

and therefore $d\hat{f}_{\varphi(p)}$ contains at least one eigenvalue in the open right-half complex plane. □

Lemma A.3 *For the dynamical system Σ, suppose $\Gamma \subset \mathcal{X}$ is a closed embedded submanifold of $\mathcal{X}$ which is globally asymptotically stable. If $N \subset \Gamma$ is a set of Lebesgue measure zero in Γ, then $\Gamma\backslash N$ is globally asymptotically stable.*

Proof To show that $\Gamma\backslash N$ is stable, it needs to be shown that for any $\varepsilon > 0$, there exists a neighborhood $\mathcal{N}(\Gamma\backslash N)$ such that, for all $x_0 \in \mathcal{N}(\Gamma\backslash N)$, $\phi(t, x_0) \in B_\varepsilon(\Gamma\backslash N)$, for all $t > 0$. Since Γ is globally asymptotically stable, there exists a neighborhood $\mathcal{N}_1(\Gamma)$ such that, for all $x_0 \in \mathcal{N}_1(\Gamma)$, $\phi(t, x_0) \in B_{\varepsilon/2}(\Gamma)$, for all $t > 0$. Since N is Lebesgue measure zero in Γ, $\Gamma\backslash N$ is dense in Γ by Proposition 6.8 in [7] and for all $\bar{x} \in N$ there exists a point $x \in \Gamma\backslash N$ such that $x \in B_{\epsilon/2}(\bar{x})$. Stability follows by choosing $\mathcal{N}(\Gamma\backslash N) = \mathcal{N}_1(\Gamma)$.

Analogously, global attractivity of $\Gamma\backslash N$ follows from global attractivity of Γ and density of $\Gamma\backslash N$ in Γ. □

The proof of Theorem A.6 is presented below.

Proof of Theorem A.6 By Lemma A.2, the isolated equilibria in $\mathcal{A}$, which are exponentially unstable equilibria relative to Γ_2, are also exponentially unstable relative to $\mathcal{X}$. It holds from Theorem 2.37 in [11] that since $\mathcal{A}$ is compact, if $\mathcal{A}$ were an infinite subset of Γ_2, then $\mathcal{A}$ would necessarily have a limit point $x \in \Gamma_2$. Since $\mathcal{A}$ is closed, it contains all its limit points and therefore $x \in \mathcal{A}$. But this contradicts that all points in $\mathcal{A}$ are isolated and therefore $\mathcal{A}$ must be finite. Suppose there are m equilibria in $\mathcal{A}$ labelled $\{x_1, \ldots, x_m\}$. Then for all $i \in \{1, \ldots, m\}$, there exists $\epsilon_i > 0$ such that $B_{\epsilon_i}(x_i) \cap \mathcal{A} = x_i$ and the minimum $\epsilon = \min\{\epsilon_1, \ldots, \epsilon_m\}$ exists. It follows that for all $i \in \{1, \ldots, m\}$, $B_\epsilon(x_i) \cap \mathcal{A} = x_i$. The condition $D(\mathcal{A}) = \bigcup_{z \in \mathcal{A}} D(\{z\})$ in Theorem A.4 holds if any solution $\phi(t, x_0)$ with initial condition $x_0 \in \mathcal{X}$ that converges to $\mathcal{A}$, necessarily converges to a particular point $x \in \mathcal{A}$. This must be the case since for any solution $\phi(t, x_0)$ converging to $\mathcal{A}$ there exists a time $T > 0$ such that $\phi(t, x_0) \in B_{\epsilon/2}(\mathcal{A})$ for all $t > T$. Therefore, there exists a point $x \in \mathcal{A}$ such that $\phi(T, x_0) \in B_{\epsilon/2}(x)$. For $t > T$ it is impossible for the solution to leave the vicinity of x and converge to another point in $\mathcal{A}$ located at least a distance ϵ away. Therefore the condition in (A.5) holds and $D(\mathcal{A})$ has zero measure by Theorem A.4. It holds from analogous arguments that the set $D(\mathcal{A}) \cap \Gamma_2$ is Lebesgue measure zero in Γ_2.

Denote $W := \mathcal{X} \backslash D(\mathcal{A})$ which is a positively invariant domain of full measure and define $\hat{\Gamma}_2 := W \cap \Gamma_2$, a set of full measure in Γ_2. Then $\hat{\Gamma}_2$ is GAS relative to $\mathcal{X}$ by Lemma A.3. Since W is positively invariant, it follows that $\hat{\Gamma}_2$ is GAS relative to W. Since $\mathcal{K}$ is stable relative to Γ_2 and W is positively invariant, $\mathcal{K}$ is also stable relative to $\hat{\Gamma}_2 = W \cap \Gamma_2$ and since $\Gamma_1 = \mathcal{A} \cup \mathcal{K}$ is globally attractive relative to Γ_2, it immediately holds that $\mathcal{K}$ is globally attractive relative to $\hat{\Gamma}_2$. Therefore $\mathcal{K}$ is GAS relative to $\hat{\Gamma}_2$ and $\hat{\Gamma}_2$ is GAS relative to W implying, by Theorem A.5, that $\mathcal{K}$ is globally asymptotically stable relative to W. This implies that $\mathcal{K}$ is almost globally asymptotically stable relative to $\mathcal{X}$. □

Appendix B
Notions of Graph Theory

Since[5] graph theory is such a broad field, in this appendix we will limit the discussion solely to notions in graph theory that are pertinent to this book. For more in depth detail on the notions introduced in this section and graph theory in general, we refer the reader to [12].

B.0.1 Basic Definitions in Graph Theory

We denote a *graph* by the pair $\mathcal{G} = (\mathcal{V}, \mathcal{E})$, where $\mathcal{V}$ is a set of nodes labelled as $\{1, \ldots, n\}$ and $\mathcal{E} = \{(i, j) \in \mathcal{V} \times \mathcal{V} : i \text{ connects to } j\}$ is the set of edges such that $(i, j) \in \mathcal{E}$ if node i connects to node j. A *directed graph* or digraph (*undirected graph*), is a graph where the edges are ordered (unordered) pairs in which $(i, j) \neq (j, i)$ $((i, j) = (j, i))$. An example of an undirected and directed graph are illustrated in Fig. B.1. Edges in a directed graph are indicated with arrows whereas edges in an undirected graph are indicated with lines. The set of *neighbors* of node i is the set $\mathcal{N}_i := \{j \in \mathcal{V} : (i, j) \in \mathcal{E}\}$. The neighbor sets in Fig. B.1 are listed in Table B.1:

Given positive numbers $a_{ij} > 0$, $i \in \mathbf{n}$, $j \in \mathcal{N}_i$, the associated *weighted Laplacian matrix* of $\mathcal{G}$ is the matrix $L = D - A$, where the i-th diagonal entry of D is the sum $\sum_{j \in \mathcal{N}_i} a_{ij}$, and A is the matrix whose element A_{ij} is a_{ij} if $j \in \mathcal{N}_i$, and 0 otherwise. The Laplacian matrix for the directed and undirected graphs in Fig. B.1 for unitary weights, i.e., $a_{ij} = 1$ for all $j \in \mathcal{N}_i$, are given by

$$L = \begin{bmatrix} 1 & -1 & 0 & 0 & 0 \\ 0 & 1 & -1 & 0 & 0 \\ 0 & 0 & 2 & -1 & -1 \\ 0 & 0 & 0 & 1 & -1 \\ -1 & -1 & -1 & 0 & 3 \end{bmatrix}, \; L = \begin{bmatrix} 2 & -1 & 0 & 0 & -1 \\ -1 & 3 & -1 & 0 & -1 \\ 0 & -1 & 3 & -1 & -1 \\ 0 & 0 & -1 & 2 & -1 \\ -1 & -1 & -1 & -1 & 4 \end{bmatrix},$$

[5] ©2018 IEEE. Reprinted, with permission, from [1]. Reused text from Sect. B.0.3.

A. Roza et al., *Distributed Coordination Theory for Robot Teams*, Lecture Notes in Control and Information Sciences 490,
https://doi.org/10.1007/978-3-030-96087-2

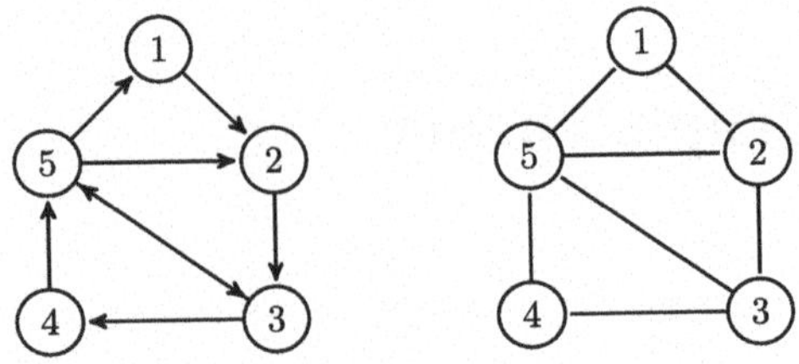

Fig. B.1 Example of a directed (left) and an undirected (right) graph

Table B.1 Neighbor sets in Fig. B.1

Node i	$\mathcal{N}_i$-Directed graph	$\mathcal{N}_i$-Undirected graph
1	{2}	{2, 5}
2	{3}	{1, 3, 5}
3	{4, 5}	{2, 4, 5}
4	{5}	{3, 5}
5	{1, 2, 3}	{1, 2, 3, 4}

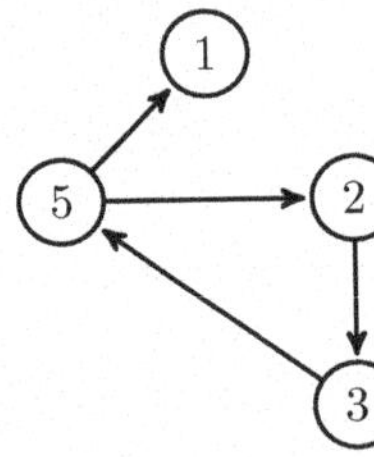

Fig. B.2 Example of a subgraph $\mathcal{G}' = (\mathcal{V}', \mathcal{E}')$ of the directed graph in Fig. B.1 with $\mathcal{V}' = \{1, 2, 3, 5\}$ and $\mathcal{E}' = \{(5,1),(5,2),(2,3),(3,5)\}$

respectively.

A graph $\mathcal{G}' = (\mathcal{V}', \mathcal{E}')$ is a *subgraph* of $\mathcal{G}$ if $\mathcal{V}' \subset \mathcal{V}$ and $\mathcal{E}' \subset \mathcal{E}$. An example of a subgraph of the directed graph $\mathcal{G}$ in Fig. B.1 is shown in Fig. B.2.

A graph $\mathcal{G}$ is called *static* if it is constant for all time and *time varying*, denoted $\mathcal{G}(t)$, if the set of nodes and/or edges varies with time t.

B.0.2 Classes of Graphs

In this section, we present several important classes of graphs. An undirected graph $\mathcal{G} = (\mathcal{V}, \mathcal{E})$ is said to be *connected* if for any two nodes $i, j \in \mathcal{V}$ there exists a path from node i to node j by traversing the edges in the graph. A *complete graph* is an undirected graph with an edge between every pair of nodes, that is, for any $i, j \in \mathcal{V}$, $(i, j) \in \mathcal{E}$. In the case of directed graphs, a graph is *strongly connected* if for any two nodes $i, j \in \mathcal{V}$ there exists a path from i to j. The undirected graph in Fig. B.1 is connected while the directed graph is strongly connected. By removing the edge $(5, 1)$, the directed graph in Fig. B.1 is no longer strongly connected.

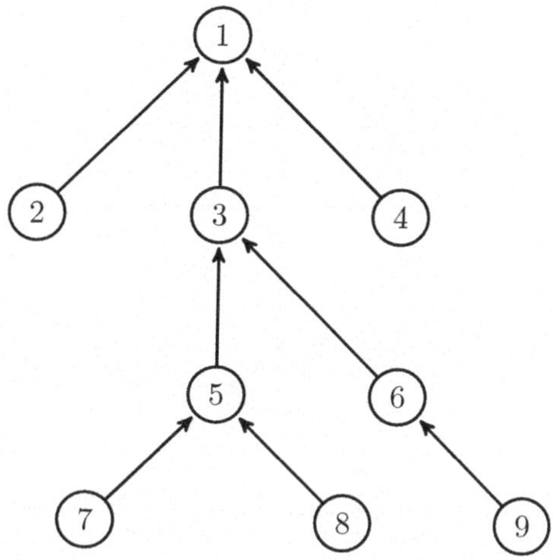

Fig. B.3 Reverse directed spanning tree with root node 1

A *directed spanning tree* is a digraph consisting of $n-1$ edges such that there exists a unique directed path from a node, called the root, to every other node. A *reverse directed spanning tree* is a graph which becomes a directed spanning tree by reversing the directions of all its edges. We identify the root of a reverse spanning tree with the root of its associated spanning tree. Therefore, in a reverse directed spanning tree the root node can be sensed indirectly by all other nodes in the graph by means of directed paths. In this sense, the root node can propagate information indirectly through the entire graph. An example of a reverse directed spanning tree is illustrated in Fig. B.3.

A digraph $\mathcal{G}$ *contains a reverse directed spanning tree* if it has a subgraph $\mathcal{G}' = (\mathcal{V}', \mathcal{E}')$ satisfying $\mathcal{V}' = \mathcal{V}$ which is a reverse directed spanning tree. A node is *globally reachable* if there exists a path from any other node to it. For a digraph $\mathcal{G}$, existence of a globally reachable node is equivalent to having a directed spanning tree in the reverse graph. An example of such a digraph is illustrated in Fig. B.4 where node 1 is a globally reachable node. Strongly connected graphs necessarily contain a globally reachable node. In fact, every node is globally reachable by definition.

If a digraph does not contain a reverse directed spanning tree, then there does not exist any node in the graph that can propagate information to the entire network. This can be interpreted as a lack of connectivity in the graph. In fact, Proposition B.1 relates the presence of a globally reachable node directly to the eigenvalues of the corresponding weighted Laplacian.

Proposition B.1 *([13, 14]) The following conditions are equivalent for a digraph* $\mathcal{G}$:

(i) $\mathcal{G}$ *contains a globally reachable node;*
(ii) For any set of positive gains $a_{ij} > 0$, $i, j \in \{1, \ldots, n\}$ *the associated weighted Laplacian matrix* L *of* $\mathcal{G}$ *is positive semi-definite, has rank* $n-1$, *and* $\operatorname{Ker} L = \operatorname{span}\{\mathbf{1}\}$.

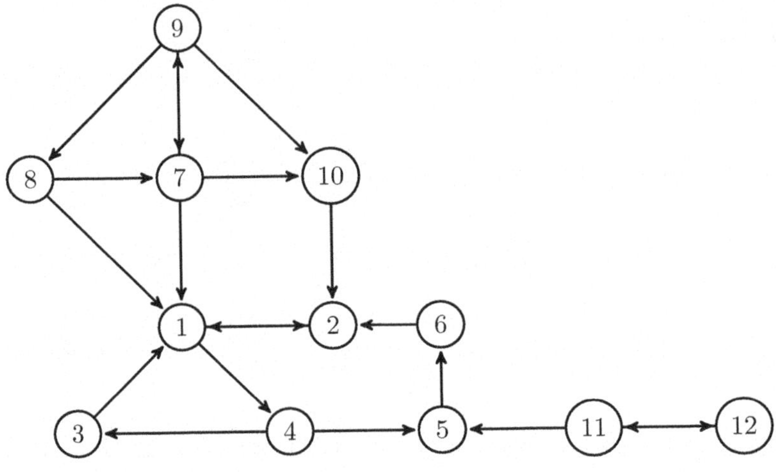

Fig. B.4 Directed graph $\mathcal{G}$ containing a reverse directed spanning tree and globally reachable node, node 1

A hierarchical digraph, defined in Definition B.1, is a class of digraphs that contains a globally reachable node.

Definition B.1 A digraph $\mathcal{G}$ with n nodes containing a globally reachable node is called a *hierarchical digraph* with $m \in \mathbb{N}$ layers if there exists a partition of $\mathcal{V}$, $\mathcal{P} = \{\mathcal{L}_\ell\}_{\ell \in \{1,\ldots,m\}}$ into m nonempty subsets called *layers* such that for all $\ell \in \{1, \ldots, m\}$ and for all $i \in \mathcal{L}_\ell$, if $(i, j) \in \mathcal{E}$ is an edge from node i to node j then $j \in \bigcup_{k \in \{1,\ldots,\ell-1\}} \mathcal{L}_k$.

In a hierarchical digraph $\mathcal{G}$, the nodes are divided into layers in which layer 1 corresponds to a single root. Nodes in layer i have neighbors only in layers $j < i$ as illustrated in Fig. B.5. As such, the root node is globally reachable and has no neighbors. As an example, this type of graph can model the behaviour seen in flocking birds that only observe other birds in front of themselves in a flocking formation.

B.0.3 Graph Decomposition

In this section, we show how a digraph containing a globally reachable node can be decomposed into so-called strongly connected components. The strongly connected components can be treated as nodes in a higher level graph known as a condensation digraph. We now give a number of formal definitions.

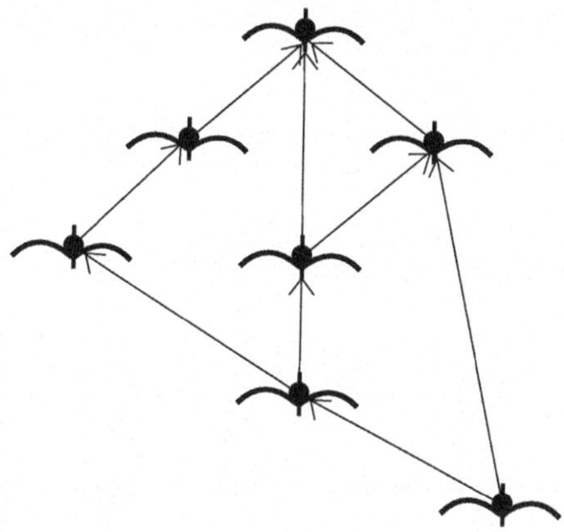

Fig. B.5 Hierarchical sensing graph

A set of nodes $S \subset \mathcal{V}$ is an *isolated component* if it has no outgoing edges, i.e., for any edge $(i, j) \in \mathcal{E}$, if $i \in S$ then $j \in S$. A subgraph $\mathcal{G}'$ is an *induced subgraph* of $\mathcal{G}$ if for any two vertices $i, j \in \mathcal{V}'$, $(i, j) \in \mathcal{E}'$ if $(i, j) \in \mathcal{E}$. A *strongly connected component* $\mathcal{G}'$ of $\mathcal{G}$ is a maximal strongly connected induced subgraph of $\mathcal{G}$. In other words, there does not exist any other strongly connected induced subgraph of $\mathcal{G}$ containing $\mathcal{G}'$. Letting $\mathcal{G}_0 = (\mathcal{V}_0, \mathcal{E}_0), \ldots, \mathcal{G}_r = (\mathcal{V}_r, \mathcal{E}_r)$ be the strongly connected components of $\mathcal{G}$, the *condensation digraph* of $\mathcal{G}$, denoted $\mathcal{C}(\mathcal{G}) = (\mathcal{V}_\mathcal{C}(\mathcal{G}), \mathcal{E}_\mathcal{C}(\mathcal{G}))$, is defined as follows. The *vertex set* $\mathcal{V}_\mathcal{C}(\mathcal{G})$ is the set of nodes $\{v_i\}_{i \in \{0,\ldots,r\}}$ where the node v_i is a *contraction* of the vertex set $\mathcal{V}_i$ of the i-th strongly connected component $\mathcal{G}_i$. The *edge set* $\mathcal{E}_\mathcal{C}(\mathcal{G})$ contains an edge (v_i, v_j) if there exist vertices $i' \in \mathcal{V}_i$ and $j' \in \mathcal{V}_j$ such that $(i', j') \in \mathcal{E}$. The following properties of the condensation digraph are found in [15].

Proposition B.2 ([15]) *Consider a digraph $\mathcal{G}$ containing a globally reachable node. The condensation $\mathcal{C}(\mathcal{G})$ satisfies the following properties:*

(i) $\mathcal{C}(\mathcal{G})$ is acyclic, i.e., there is no path in $\mathcal{C}(\mathcal{G})$ beginning and ending at the same node;
(ii) $\mathcal{C}(\mathcal{G})$ contains a reverse directed spanning tree $\mathcal{T}$ with a unique root $v_0 \in \mathcal{V}_\mathcal{C}(\mathcal{G})$;
(iii) There exists at least one vertex $v_i \in \mathcal{V}_\mathcal{C}(\mathcal{G})$ such that v_0 is the only neighbor of v_i.

An example of a digraph $\mathcal{G}$ containing a reverse directed spanning tree is shown in Fig. B.6. The strongly connected components are boxed. The resulting acyclic condensation digraph $\mathcal{C}(\mathcal{G})$ is shown in Fig. B.7. The vertex v_0 in the figure is the unique root of the reverse directed spanning tree in $\mathcal{C}(\mathcal{G})$.

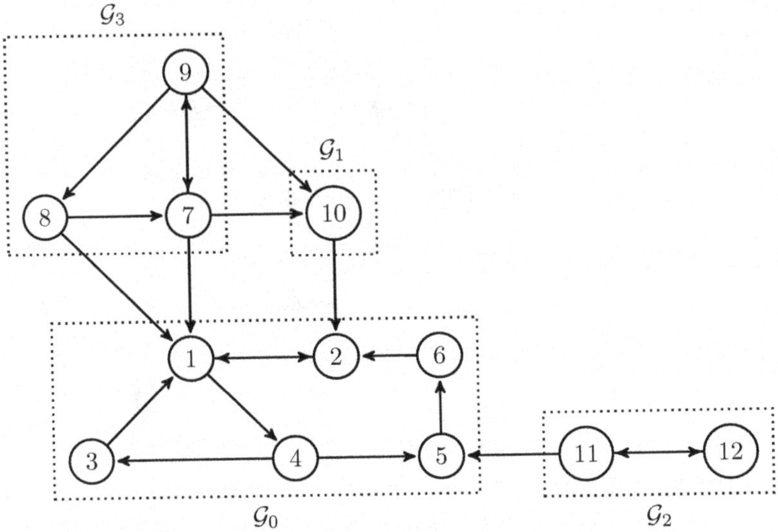

Fig. B.6 Directed graph $\mathcal{G}$ containing a reverse directed spanning tree. The strongly connected components $\mathcal{G}_0, \ldots, \mathcal{G}_3$ are boxed

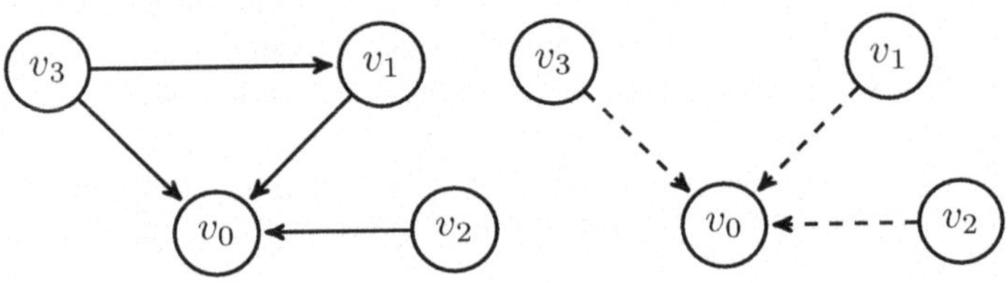

Fig. B.7 Condensation digraph $\mathcal{C}(\mathcal{G})$ associated with the graph $\mathcal{G}$ in Fig. B.6 (left) and reverse directed spanning tree contained in $\mathcal{C}(\mathcal{G})$ (right)

As in [15], we define the vertex set $\mathcal{L}_j \subset \mathcal{V}$ to be the union of those vertex sets $\mathcal{V}_i$ that correspond to vertices v_i in the condensation digraph with the property that the maximal path length from v_i to the root v_0 is equal to j. By this definition, $\mathcal{L}_0 := \mathcal{V}_0$. We let $\mathcal{L}_{-1} := \varnothing$. Defining the vertex set $\bar{\mathcal{L}}_j := \cup_{i=0}^{j} \mathcal{L}_i$, by construction, the neighbors of any vertex in $\mathcal{L}_j$ are contained in $\bar{\mathcal{L}}_{j-1}$. Therefore each node set $\bar{\mathcal{L}}_j$ is isolated. For the example in Fig. B.7, we have $\mathcal{L}_0 = \{1, 2, 3, 4, 5, 6\}$, $\mathcal{L}_1 = \{10\} \cup \{11, 12\}$, and $\mathcal{L}_2 = \{7, 8, 9\}$.

References

1. Roza, A., Maggiore, M., Scardovi, L.: A smooth distributed feedback for global rendezvous of unicycles. IEEE Trans. Control Netw. Syst. **5**(1), 640–652 (2018)
2. Khalil, H.: Nonlinear Systems, 3 edn. Prentice Hall (2002)
3. Allen, R.: Mathematical Analysis for Economists. Macmillan and Company Limited (1938)
4. Freeman, R.: A global attractor consisting of exponentially unstable equilibria. In: American Control Conference (ACC), pp. 4855–4860 (2013)
5. Bhatia, N., Szegö, G.: Stability Theory of Dynamical Systems. Springer Science & Business Media (2002)
6. Bhat, S., Bernstein, D.: A topological obstruction to continuous global stabilization of rotational motion and the unwinding phenomenon. Syst. Control Lett. **39**(1), 63–70 (2000)
7. Lee, J.: Introduction to Smooth Manifolds, 2 edn. Springer (2013)
8. El-Hawwary, M., Maggiore, M.: Reduction theorems for stability of closed sets with application to backstepping control design. Automatica **49**(1), 214–222 (2013)
9. Seibert, P., Florio, J.: On the reduction to a subspace of stability properties of systems in metric spaces. Annali di Matematica **169**(1), 291–320 (1995)
10. Maggiore, M.: Reduction principles for hierarchical control design. Notes for a minicourse offered at the Norwegian University of Science and Technology, Trondheim (2015)
11. Rudin, W.: Principles of Mathematical Analysis, Vol. 3. McGraw-hill (1976)
12. Godsil, C., Royle, G.: Algebraic Graph Theory. Springer (2001)
13. Ren, W., Beard, R.: Consensus seeking in multiagent systems under dynamically changing interaction topologies. IEEE Trans. Autom. Control **50**(5), 655–661 (2005)
14. Lin, Z., Francis, B., Maggiore, M.: Necessary and sufficient graphical conditions for formation control of unicycles. IEEE Trans. Autom. Control **50**(1), 121–127 (2005)
15. Hatanaka, T., Chopra, N., Fujita, M., Spong, M.: Passivity-Based Control and Estimation in Networked Robotics. Springer (2015)

A. Roza et al., *Distributed Coordination Theory for Robot Teams*,
Lecture Notes in Control and Information Sciences 490,
https://doi.org/10.1007/978-3-030-96087-2

Index

A. Roza et al., *Distributed Coordination Theory for Robot Teams*,
Lecture Notes in Control and Information Sciences 490,
https://doi.org/10.1007/978-3-030-96087-2

The manufacturer's authorised representative in the EU is Springer Nature Customer Service Centre GmbH, Europaplatz 3, 69115 Heidelberg, Germany. If you have any concerns regarding our products, please contact ProductSafety@springernature.com

Printed and bound by CPI Group (UK) Ltd, Croydon, CR0 4YY
22/09/2026
02221055-0018